Kim-Kwang Raymond Choo ·
Tommy Morris · Gilbert L. Peterson ·
Eric Imsand
Editors

National Cyber Summit (NCS) Research Track 2020

Editors
Kim-Kwang Raymond Choo
Department of Information Systems
The University of Texas at San Antonio
San Antonio, TX, USA

Gilbert L. Peterson
Air Force Institute of Technology
Wright-Patterson AFB, OH, USA

Tommy Morris
The University of Alabama in Huntsville
Huntsville, AL, USA

Eric Imsand
The University of Alabama in Huntsville
Huntsville, AL, USA

ISSN 2194-5357 ISSN 2194-5365 (electronic)
Advances in Intelligent Systems and Computing
ISBN 978-3-030-58702-4 ISBN 978-3-030-58703-1 (eBook)
https://doi.org/10.1007/978-3-030-58703-1

This Springer imprint is published by the registered company Springer Nature Switzerland AG
The registered company address is: Gewerbestrasse 11, 6330 Cham, Switzerland

Preface

Cyberthreats remain important and strategically relevant in both developed and developing economies. For example, in the "Worldwide threat assessment of the US intelligence community (January 29, 2019)[1], it was reported that:

> *Our adversaries and strategic competitors will increasingly use cyber capabilities – including cyber espionage, attack, and influence—to seek political, economic, and military advantage over the United States and its allies and partners*

Hence, there is a need to keep a watchful brief on the cyberthreat landscape, particularly as technology advances and new cyberthreats emerge. This is the intention of this conference proceedings.

This conference proceedings contains 12 regular papers and 4 short papers from the 2020 National Cyber Summit Research Track. The proceedings also includes one invited paper. The 2020 National Cyber Summit was originally planned to be held in Huntsville, Alabama, from June 2 to 4, 2020. However, due to the COVID-19 pandemic, all tracks of the 2020 National Cyber Summit, except the Research Track, were delayed until June 2021. The 2020 National Cyber Summit Research Track was held via video conference on June 3 and 4. Authors from each selected paper presented their work and took questions from the audience.

The papers were selected from submissions from universities, national laboratories, and the private sector from across the USA, Bangladesh, and Slovenia. All of the papers went through an extensive review process by internationally recognized experts in cybersecurity.

The Research Track at the 2020 National Cyber Summit has been made possible by the joint effort of a large number of individuals and organizations worldwide. There is a long list of people who volunteered their time and energy to put together the conference and deserved special thanks. First and foremost, we would like to offer our gratitude to the entire Organizing Committee for guiding the entire process

[1]https://www.dni.gov/files/ODNI/documents/2019-ATA-SFR—SSCI.pdf.

of the conference. We are also deeply grateful to all the Program Committee members for their time and efforts in reading, commenting, debating, and finally selecting the papers. We also thank all the external reviewers for assisting the Program Committee in their particular areas of expertise as well as all the authors, participants, and session chairs for their valuable contributions.

Tommy Morris
Kim-Kwang Raymond Choo
General Chairs

Gilbert L. Peterson
Eric Imsand
Program Committee Chairs

Organization

Organizing Committee

General Chairs

Tommy Morris	The University of Alabama in Huntsville, USA
Kim-Kwang Raymond Choo	The University of Texas at San Antonio, USA

Program Committee Chairs

Gilbert L. Peterson	Air Force Institute of Technology, USA
Eric Imsand	The University of Alabama in Huntsville, USA

Program Committee and External Reviewers

Program Committee Members

Tommy Morris	The University of Alabama in Huntsville, USA
Kim-Kwang Raymond Choo	The University of Texas at San Antonio, USA
George Grispos	The University of Nebraska at Omaha, USA
Eric Imsand	The University of Alabama in Huntsville, USA
Nour Moustafa	The University of New South Wales at Canberra, Australia
Wei Zhang	The University of Louisville, USA
Reza M. Parizi	Kennesaw State University, USA
Octavio Loyola-González	Tecnologico de Monterrey, Mexico
Ezhil Kalaimannan	The University of West Florida, USA
Vijayan Sugumaran	Oakland University, USA
Vahid Heydari	Rowan University, USA
Jianyi Zhang	Beijing Electronic Science and Technology Institute, China
Gilbert Petersen	US Air Force Institute of Technology, USA
Ravi Rao	Fairleigh Dickinson University, USA

Jun Dai	California State University, Sacramento, USA
David Dampier	Marshall University, USA
Patrick Pape	Auburn University Montgomery, USA
Cong Pu	Marshall University, USA
Junggab Son	Kennesaw State University, USA
John Bland	The University of Alabama in Huntsville, USA
Robin Verma	The University of Texas at San Antonio, USA
Miguel Angel Medina Pérez	Tecnologico de Monterrey, Mexico
Huijun Wu	Arizona State University, USA
Rongxing Lu	The University of New Brunswick, Canada

External Reviewers

Arquímides Méndez Molina
Daniel Colvett
Duo Lu
Erdal Kose
Feng Wang
Gautam Srivastava
Hossain Shahriar
Iman Vakilinia
Irene Kasian
James Okolica
Jorge Rodriguez Ruiz
Lazaro Bustio
Lei Xu
Mariano Vargas-Santiago
Miguel Ángel Álvarez Carmona
Mohammad Shojaeshafiei
Mohiuddin Ahmed
Nickolaos Koroniotis
Ning Wang
Nisheeth Agrawal
Rachel Foster
Seyedamin Pouriyeh
Swapnoneel Roy
Tania Williams
Tymaine Whitaker
Xiaomei Zhang
Yan Huang
Yaoqing Liu
Younghun Chae
Zachary Tackett
Zhimin Gao

Contents

Invited Paper

Cyber Security Education

Cyber Security Technology

Short Papers

Invited Paper

Experiences and Lessons Learned Creating and Validating Concept Inventories for Cybersecurity

Alan T. Sherman[1(✉)], Geoffrey L. Herman[2], Linda Oliva[1], Peter A. H. Peterson[3], Enis Golaszewski[1], Seth Poulsen[2], Travis Scheponik[1], and Akshita Gorti[1]

[1] Cyber Defense Lab, University of Maryland, Baltimore County (UMBC), Baltimore, MD 21250, USA
sherman@umbc.edu

[2] Computer Science, University of Illinois at Urbana-Champaign, Urbana, IL 61801, USA
glherman@illinois.edu

[3] Department of Computer Science, University of Minnesota Duluth, Duluth, MN 55812, USA
pahp@d.umn.edu

http://www.csee.umbc.edu/people/faculty/alan-t-sherman/, http://publish.illinois.edu/glherman/, http://www.d.umn.edu/~pahp/

Abstract. We reflect on our ongoing journey in the educational *Cybersecurity Assessment Tools (CATS)* Project to create two concept inventories for cybersecurity. We identify key steps in this journey and important questions we faced. We explain the decisions we made and discuss the consequences of those decisions, highlighting what worked well and what might have gone better.

The CATS Project is creating and validating two concept inventories—conceptual tests of understanding—that can be used to measure the effectiveness of various approaches to teaching and learning cybersecurity. The *Cybersecurity Concept Inventory (CCI)* is for students who have recently completed any first course in cybersecurity; the *Cybersecurity Curriculum Assessment (CCA)* is for students who have recently completed an undergraduate major or track in cybersecurity. Each assessment tool comprises 25 *multiple-choice questions (MCQs)* of various difficulties that target the same five core concepts, but the CCA assumes greater technical background.

Key steps include defining project scope, identifying the core concepts, uncovering student misconceptions, creating scenarios, drafting question stems, developing distractor answer choices, generating educational materials, performing expert reviews, recruiting student subjects, organizing workshops, building community acceptance, forming a team and nurturing collaboration, adopting tools, and obtaining and using funding.

Creating effective MCQs is difficult and time-consuming, and cybersecurity presents special challenges. Because cybersecurity issues are often

K.-K. R. Choo et al. (Eds.): NCS 2020, AISC 1271, pp. 3–34, 2021.
https://doi.org/10.1007/978-3-030-58703-1_1

subtle, where the adversarial model and details matter greatly, it is challenging to construct MCQs for which there is exactly one best but non-obvious answer. We hope that our experiences and lessons learned may help others create more effective concept inventories and assessments in STEM.

Keywords: Computer science education · Concept inventories · Cryptography · Cybersecurity Assessment Tools (CATS) · Cybersecurity education · Cybersecurity Concept Inventory (CCI) · Cybersecurity Curriculum Assessment (CCA) · Multiple-choice questions

1 Introduction

When we started the *Cybersecurity Assessment Tools (CATS)* Project [44] in 2014, we thought that it should not be difficult to create a collection of 25 *multiple choice questions (MCQs)* that assess student understanding of core cybersecurity concepts. Six years later, in the middle of validating the two draft assessments we have produced, we now have a much greater appreciation for the significant difficulty of creating and validating effective and well-adopted concept inventories. This paper highlights and reflects on critical steps in our journey, with the hope that our experiences can provide useful lessons learned to anyone who wishes to create a cybersecurity concept inventory, any assessment in cybersecurity, or any assessment in STEM.[1]

Cybersecurity is a vital area of growing importance for national competitiveness, and there is a significant need for cybersecurity professionals [4]. The number of cybersecurity programs at colleges, universities, and training centers is increasing. As educators wrestle with this demand, there is a corresponding awareness that we lack a rigorous research base that informs how to prepare cybersecurity professionals. Existing certification exams, such as CISSP [9], are largely informational, not conceptual. We are not aware of any scientific analysis of any of these exams. Validated assessment tools are essential so that cybersecurity educators have trusted methods for discerning whether efforts to improve student preparation are successful [32]. The CATS Project provides rigorous evidence-based instruments for assessing and evaluating educational practices; in particular, they will help assess approaches to teaching and learning cybersecurity such as traditional lecture, case study, hands-on lab exercises, interactive simulation, competition, and gaming.

We have produced two draft assessments, each comprising 25 MCQs. The *Cybersecurity Concept Inventory (CCI)* measures how well students understand core concepts in cybersecurity after a first course in the field. The *Cybersecurity Curriculum Assessment (CCA)* measures how well students understand core concepts after completing a full cybersecurity curriculum. Each test item comprises a *scenario*, a *stem* (a question), and five *alternatives* (answer choices comprising

[1] Science, Technology, Engineering, and Mathematics (STEM).

a single best answer choice and four distractors). The CCI and CCA target the same five core concepts (each being an aspect of adversarial thinking), but the CCA assumes greater technical background. In each assessment, there are five test items of various difficulties for each of the five core concepts.

Since fall 2014, we have been following prescriptions of the National Research Council for developing rigorous and valid assessment tools [26,35]. We carried out two surveys using the Delphi method to identify the scope and content of the assessments [33]. Following guidelines proposed by Ericsson and Simon [13], we then carried out qualitative interviews [40,45] to develop a cognitive theory that can guide the construction of assessment questions. Based on these interviews, we have developed a preliminary battery of over 30 test items for each assessment. Each test item measures how well students understand core concepts as identified by our Delphi studies. The distractors (incorrect answers) for each test item are based in part on student misconceptions observed during the interviews. We are now validating the CCI and CCA using small-scale pilot testing, cognitive interviews, expert review, and large-scale psychometric testing [31].

The main contributions of this paper are lessons learned from our experiences with the CATS Project. These lessons include strategies for developing effective scenarios, stems, and distractors, recruiting subjects for psychometric testing, and building and nurturing effective collaborations. We offer these lessons, not with the intent of prescribing advice for all, but with the hope that others may benefit through understanding and learning from our experiences. This paper aims to be the paper we wished we could have read before starting our project.

2 Background and Previous and Related Work

We briefly review relevant background on concept inventories, cybersecurity assessments, and other related work. To our knowledge, our CCI and CCA are the first concept inventories for cybersecurity, and there is no previous paper that presents lessons learned creating and validating any concept inventory.

2.1 Concept Inventories

A *concept inventory (CI)* is an assessment (typically multiple-choice) that measures how well student conceptual knowledge aligns with accepted conceptual knowledge [23]. Concept inventories have been developed for many STEM disciplines, consistently revealing that students who succeed on traditional classroom assessments struggle to answer deeper conceptual questions [6,14,19,23,28]. When students have accurate, deep conceptual knowledge, they can learn more efficiently, and they can transfer their knowledge across contexts [28]. CIs have provided critical evidence supporting the adoption of active learning and other evidence-based practices [14,19,20,29]. For example, the *Force Concept Inventory (FCI)* by Hestenes et al. [23] “spawned a dramatic movement of reform in physics education.” [12, p. 1018].

For CIs to be effective, they need to be validated. Unfortunately, few CIs have undergone rigorous validation [34,47]. Validation is a chain of evidence that supports the claims that an assessment measures the attributes that it claims to measure. This process requires careful selection of what knowledge should be measured, carefully constructing questions that are broadly accepted as measuring that knowledge, and providing statistical evidence that the assessment is internally consistent. The usefulness of a CI is threatened if it fails any of these requirements. Additionally, a CI must be easy to administer, and its results must be easy to interpret—or they can easily be misused. Critically, CIs are intended as research instruments that help instructors make evidence-based decisions to improve their teaching and generally should not be used primarily to assign student grades or to evaluate a teacher's effectiveness.

Few validated CIs have been developed for computing topics; notable exceptions include the Digital Logic Concept Inventory [22] (led by CATS team member Herman) and early work on the Basic Data Structures Inventory [37]. None has been developed for security related topics.

2.2 Cybersecurity Assessment Exams

We are not aware of any other group that is developing an educational assessment tool for cybersecurity. There are several existing certification exams, including ones listed by NICCS as relevant [11].

CASP+ [8] comprises multiple-choice and performance tasks items including enterprise security, risk management, and incident response. OSCP [41] (offensive security) is a 24-hour practical test focusing on penetration testing. Other exams include CISSP, Security+, and CEH [7,9,46], which are mostly informational, not conceptual. Global Information Assurance Certification (GIAC) [10] offers a variety of vendor-neutral MCQ certification exams linked to SANS courses; for each exam type, the gold level requires a research paper. We are unaware of any scientific study that characterizes the properties of any of these tests.

2.3 Other Related Work

The 2013 IEEE/ACM Computing Curriculum Review [25] approached the analysis of cybersecurity content in undergraduate education from the perspective of traditional university curriculum development. Later, the ACM/IEEE/AIS SIGSEC/IFIP Joint Task Force on Cybersecurity Education (JTF) [15] developed comprehensive curricular guidance in cybersecurity education, releasing Version 1.0 of their guidelines at the end of 2017.

To improve cybersecurity education and research, the National Security Agency (NSA) and Department of Homeland Security (DHS) jointly sponsor the National Centers of Academic Excellence (CAE) program. Since 1998, more than 300 schools have been designated as CAEs in Cyber Defense. The requirements include sufficiently covering certain "Knowledge Units" (KUs) in their academic programs, making the CAE program a "significant influence on the curricula of programs offering cybersecurity education" [16].

The NICE Cybersecurity Workforce Framework [30] establishes a common lexicon for explaining a structured description of professional cybersecurity positions in the workforce with detailed documentation of the knowledge, skills, and abilities needed for various types of cybersecurity activities.

More recently, the Accreditation Board of Engineering and Technology (ABET) has included, in the 2019–2020 Criteria for Accrediting Computing Program, criteria for undergraduate cybersecurity (or similarly named) programs. ABET has taken an approach similar to that of the CAE program, requiring coverage of a set of topics without requiring any specific set of courses.

In a separate project, CATS team member Peterson and his students [24,36] worked with experts to identify specific and persistent commonsense misconceptions in cybersecurity, such as that “physical security is not important,” or that “encryption is a foolproof security solution.” They are developing a CI focusing on those misconceptions.

3 Key Steps and Takeaways from the CATS Project

We identify the key steps in our journey creating and validating the CCI and CCA. For each step, we comment on important issues, decisions we made, the consequences of those decisions, and the lessons we learned.

3.1 Genesis of the CATS Project

On February 24–25, 2014, Sherman, an expert in cybersecurity, participated in a NSF workshop to advise NSF on how to advance cybersecurity education. NSF occasionally holds such workshops in various areas and distributes their reports, which can be very useful in choosing research projects. The workshop produced a list of prioritized recommendations, beginning with the creation of a concept inventory [5]. At the workshop, Sherman met one of his former MIT officemates, Michael Loui. Sherman proposed to Loui that they work together to create such a concept inventory. About to retire, Loui declined, and introduced Sherman to Loui's recent PhD graduate Herman, an expert in engineering education. Without meeting in person for over a year, Sherman and Herman began a productive collaboration. Loui's introduction helped establish initial mutual trust between Sherman and Herman.

3.2 Defining Scope of Project and Assessment Tools

As with many projects, defining scope was one of the most critical decisions of the CATS Project. We pondered the following questions, each of whose answers had profound implications on the direction and difficulty of the project. How many assessment tools should we develop? For what purposes and subject populations should they be developed? In what domain should the test items be cast? Should the test items be MCQs?

We decided on creating two tools: the CCI (for students in any first course in cybersecurity) and CCA (for recent graduates of a major or track in cybersecurity), because there is a strong need for each, and each tool has different requirements. This decision doubled our work. Creating any more tools would have been too much work.

Our driving purpose is to measure the effectiveness of various approaches to teaching cybersecurity, not to evaluate the student or the instructor. This purpose removes the need for a high-stakes test that would require substantial security and new questions for each test instance. By contrast, many employers who have talked with us about our work have stated their desire for an instrument that would help them select whom to hire (our assessments are neither designed nor validated for that high-stakes purpose).

Ultimately, we decided that all test items should be cast in the domain of cybersystems, on the grounds that cybersecurity takes place in the context of such systems. Initially, however, we experimented with developing test items that probed security concepts more generally, setting them in a variety of every-day contexts, such as building security, transportation security, and physical mail. Both approaches have merit but serve different purposes.

Following the format of most concept inventories, we decided that each test item be a MCQ. For more about MCQs and our reasons for using them, see Sect. 4.

3.3 Identifying Core Concepts

The first major step in creating any concept inventory is to identify the core concepts to be tested. We sought about five important, difficult, timeless, cross-cutting concepts. These concepts do not have to cover cybersecurity comprehensively. For example, the Force Concept Inventory targets five concepts from Newtonian dynamics, not all concepts from physics. To this end, we engaged 36 cybersecurity experts in two Delphi processes, one for the CCI and one for the CCA [33]. A Delphi process is a structured process for achieving consensus on contentious issues [3,18].

An alternative to the Delphi process is the focus group. Although focus groups can stimulate discussions, they can be influenced strongly by personalities and it can be difficult to organize the results coherently. For example, attempts to create concept maps for cybersecurity via focus groups have struggled to find useful meaning in the resulting complex maps, due to their high density.[2]

Delphi processes also have their challenges, including recruiting and retaining experts, keeping the experts focused on the mission, and processing expert comments, including appropriately grouping similar comments. We started with 36 experts in total, 33 for CCI, 31 for CCA, and 29 in both. We communicated with the experts via email and SurveyMonkey. For each process, approximately 20 experts sustained their efforts throughout. Many of the experts came with

[2] Personal correspondence with Melissa Dark (Purdue).

strongly held opinions to include their favorite topics, such as policy, forensics, malware analysis, and economic and legal aspects. We completed the two Delphi processes in parallel in fall 2014, taking about eight weeks, conducting initial topic identification followed by three rounds of topic ratings. Graduate research assistant Parekh helped orchestrate the processes. It is difficult to recruit and retain experts, and it is a lot of work to process the large volume of free-form comments.

The first round produced very similar results for both Delphi processes, with both groups strongly identifying aspects of adversarial thinking. Therefore, we restarted the CCI process with an explicit focus on adversarial thinking. After each round, using principles of grounded theory [17], we grouped similar responses and asked each expert to rate each response on a scale from one to ten for importance and timeliness. We also encouraged experts to explain their ratings. We communicated these ratings and comments (without attribution) to everyone. The CCA process produced a long list of topics, with the highest-rated ones embodying aspects of adversarial thinking.

In the end, the experts came to a consensus on five important core concepts, which deal with adversarial reasoning (see Table 1). We decided that each of the two assessment tools would target these same five concepts, but assume different levels of technical depth.

Table 1. The five core concepts underlying the CCI and CCA embody aspects of adversarial thinking.

1	Identify vulnerabilities and failures
2	Identify attacks against CIA triad[a] and authentication
3	Devise a defense
4	Identify the security goals
5	Identify potential targets and attackers

[a]CIA Triad (Confidentiality, Integrity, Availability).

3.4 Interviewing Students

We conducted two types of student interviews: talk-aloud interviews to uncover student misconceptions [45], and cognitive interviews as part of the validation process [31]. We conducted the interviews with students from three diverse schools: UMBC (a public research university), Prince George's Community College, and Bowie University (a Historically Black College or University (HBCU)). UMBC's Institutional Review Board (IRB) approved the protocol.

We developed a series of scenarios based on the five core concepts identified in the Delphi processes. Before drafting complete test items, we conducted 26 one-hour talk-aloud interviews to uncover misconceptions, which we subsequently used to generate distractors. During the interviews we asked open-ended questions of various difficulties based on prepared scenarios. For each scenario, we also prepared a "tree" of possible hints and follow-up questions, based on the student's progress. The interviewer explained that they wanted to understand how the student thought about the problems, pointing out that the interviewer was not an expert in cybersecurity and that they were not evaluating the student. One or two cybersecurity experts listened to each interview, but reserved any possible comments or questions until the end. We video- and audio-recorded each interview.

We transcribed each interview and analyzed it using novice-led paired thematic analysis [45]. Labeling each section of each interview as either "correct" or "incorrect," we analyzed the data for patterns of misconceptions. Four themes emerged: overgeneralizations, conflated concepts, biases, and incorrect assumptions [45]. Together, these themes reveal that students generally failed to grasp the complexity and subtlety of possible vulnerabilities, threats, risks, and mitigations.

As part of our validation studies, we engaged students in cognitive interviews during which a student reasoned aloud while they took the CCI or CCA. These interviews helped us determine if students understood the questions, if they selected the correct answer for the correct reason, and if they selected incorrect answers for reasons we had expected. These interviews had limited contributions since most subjects had difficulty providing rationales for their answer choices. The interviews did reveal that specific subjects had difficulty with some of the vocabulary, prompting us to define selected terms (e.g., masquerade) at the bottom of certain test items.

Although there is significant value in conducting these interviews, they are a lot of work, especially analysis of the talk-aloud interviews. For the purpose of generating distractors, we now recommend very strongly the simpler technique of asking students (including through crowdsourcing) open-ended stems, without providing any alternatives (see Sect. 3.7).

3.5 Creating Scenarios

To prepare for our initial set of interviews (to uncover student misconceptions), we created several interview prompts, each based on an engaging scenario. Initially we created twelve scenarios organized in three sets of four, each set including a variety of settings and difficulty levels.

We based our first CCI test items on the initial twelve scenarios, each test item comprising a scenario, stem, and five answer choices. Whenever possible, to keep the stem as simple as possible, we placed details in the scenario rather than in the stem. Initially, we had planned to create several stems for each scenario, but as we explain in Sect. 4, often this plan was hard to achieve. Over time, we created many more scenarios, often drawing from our life experiences. Sometimes we would create a scenario specifically intended to target a specific concept (e.g., identify the attacker) or topic (e.g., cyberphysical system).

For example, one of our favorite CCI scenarios is a deceptively simple one based on lost luggage. We created this scenario to explore the concept of identifying targets and attackers.

Lost Luggage. *Bob's manager Alice is traveling abroad to give a sales presentation about an important new product. Bob receives an email with the following message: "Bob, I just arrived and the airline lost my luggage. Would you please send me the technical specifications? Thanks, Alice."*

Student responses revealed a dramatic range of awareness and understanding of core cybersecurity concepts. Some students demonstrated lack of adversarial thinking in suggesting that Bob should simply e-mail the information to Alice, reflecting lack of awareness of potential threats, such as someone impersonating Alice or eavesdropping on the e-mail. Similarly, others recognized the need to authenticate Alice, but still recommended e-mailing the information without encryption after authenticating Alice. A few students gave detailed thoughtful answers that addressed a variety of concerns including authentication, confidentiality, integrity, policy, education, usability, and best practices.

We designed the CCA for subjects with greater technical sophistication, for which scenarios often include an artifact (e.g., program, protocol, log file, system diagram, or product specification). We based some CCA test items directly on CCI items, adding an artifact. In most cases we created entirely new scenarios. In comparison with most CCI scenarios, CCA scenarios with artifacts require students to reason about more complex real-world challenges in context with specific technical details, including ones from the artifact. For example, inspired by a network encountered by one of our team members, the CCA switchbox scenario (Fig. 1) describes a corporate network with switchbox. We present this scenario using prose and a system diagram and use it to target the concept of identifying security goals. As revealed in our cognitive interviews, these artifacts inspired and challenged students to apply concepts to complex situations. A difficulty in adding artifacts is to maintain focus on important timeless concepts and to minimize emphasizing particular time-limited facts, languages, or conventions.

Responses from our new crowdsourcing experiment (Sect. A) suggest that some subjects were confused about how many LANs could be connected through the switch simultaneously. Consequently, we made one minor clarifying edit to the last sentence of the scenario: we changed "switch that physically connects the computer to the selected LAN" to "switch that physically connects the computer to exactly one LAN at a time."

Switchbox. *A company has two internal Local Area Networks (LANs): a core LAN connected to an email server and the Internet, and an accounting LAN connected to the corporate accounting server (which is not connected to the Internet). Each desktop computer has one network interface card. Some computers are connected to only one of the networks (e.g., Computers A and C). A computer that requires access to both LANs (e.g., Computer B) is connected to a switchbox with a toggle switch that physically connects the computer to exactly one LAN at a time.*

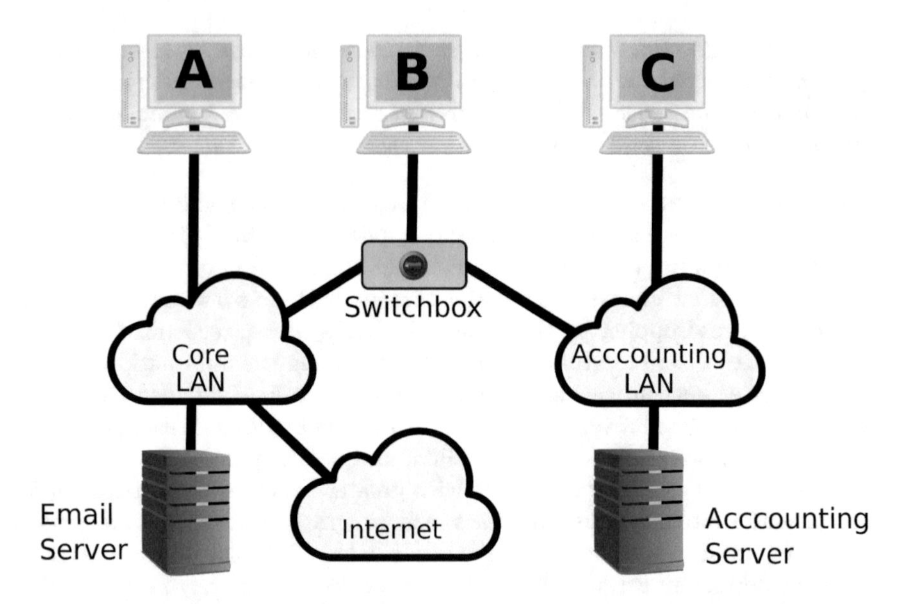

Fig. 1. CCA switchbox scenario, which includes an artifact of a network diagram with switchbox.

Comparing the CCI "lost luggage" scenario to the CCA "switchbox" scenario, one can see that the CCI scenario is simple, requiring few details to be clear. On the other hand, the CCA scenario requires the Consideration and analysis of a greater number of facts and properties of the system. Some of these facts, such as "[the accounting LAN] is not connected to the Internet" and that "each desktop computer has one network interface card," may have been added in discussion as the problem developers required clarification in their discussion of the scenario. In conjunction with the artifact, the scenario serves to constrain the problem in such a way that the system can be well-understood.

3.6 Drafting Stems

Drafting a stem requires careful consideration of several points, in the context of the scenario and alternatives. Each test item should primarily target one of the five core concepts, though to some degree it might involve additional concepts. The stem should be meaningful by itself, and an expert should be able to answer it even without being provided any of the alternatives. We try to to keep each stem as focused and short as reasonably possible. To this end, we try to place most of the details into the scenario, though stems may add a few supplemental details. Each test item should measure conceptual understanding, not informational knowledge, and not intelligence. Throughout we consider the "Vanderbilt" guidelines [2], which, among other considerations, caution against negatively worded stems, unless some significant learning outcome depends on such negativity (e.g., "What should you **NOT** use to extinguish an oil fire?").

There are many pitfalls to avoid, including unclear wording, ambiguity, admitting multiple alternatives, using unfamiliar words, and being too easy or too hard. As a rule of thumb, to yield useful information, the difficulty of each test item should be set so that at least 10%, and at most 90%, of the subjects answer correctly. We try hard to leave nothing to the subject's imagination, making it a mistake for subjects to add details or assumptions of their own creation that are not explicitly in the scenario or stem.

To carry out the detailed work of crafting test items, we created a problem development group, whose regular initial members were cybersecurity experts Sherman, Golaszewski, and Scheponik. In fall 2018, Peterson joined the group. Often during our weekly CATS conference calls, we would present a new CCI item to Herman and Oliva, who are not cybersecurity experts. It was helpful to hear the reactions of someone reading the item for the first time and of someone who knows little about cybersecurity. An expert in MCQs, Herman was especially helpful in identifying unintentional clues in the item. Herman and Oliva were less useful in reviewing the more technical CCA test items.

We created and refined stems in a highly iterative process. Before each meeting, one member of the problem development group would prepare an idea, based strongly on one of our core concepts. During the meeting, this member would present their suggestion through a shared Google Doc, triggering a lively discussion. Other members of the group would raise objections and offer possible improvements, while simultaneously editing the shared document. Having exactly three or four members present worked extremely well for us, to provide the diverse perspectives necessary to identify and correct issues, while keeping the discussion controlled enough to avoid anarchy and to permit everyone to engage. Over time we became more efficient and skilled at crafting test items, because we could better overcome predictable difficulties and avoid common missteps.

Sometimes, especially after receiving feedback from students or experts, we would reexamine a previously drafted test item. Having a fresh look after the passage of several weeks often helped us to see new issues and improvements.

Continuing the switchbox example from Sect. 3.5, Fig. 2 gives three versions of this CCA stem during its evolution. In Version 1, we deemed the open-ended phrasing as too subjective since it is impossible for the subject to determine definitely the network design's primary intent. This type of open-ended stem risks leading to multiple acceptable alternatives, or to one obviously correct alternative and four easily rejected distractors. Neither of these outcomes would be acceptable.

In Version 2, instead of asking about the designer's intent, we ask about security goals that this design supports. We also settled on a unified language for stems, using the verb "choose," which assertively emphasizes that the subject should select one best answer from the available choices. This careful wording permits the possibility that the design might support multiple security goals, while encouraging one of the security goals to be more strongly supported than the others.

Seeking even greater clarity and less possible debate over what is the best answer, we iterated one more time. In Version 3, we move away from the possibly

Version 1: What security goal is this design primarily intended to meet?
Version 2: Choose the security goal that this design best supports.
Version 3: Choose the action that this design best prevents.

Fig. 2. Evolution of the stem for the CCA switchbox test item.

subjective phrase "goal that this design best supports" and instead focus on the more concrete "action that this design best prevents." This new wording also solves another issue: because the given design is poor, we did not wish to encourage subjects to think that we were praising the design. This example illustrates the lengthy, careful, detailed deliberations we carried out to create and refine stems.

3.7 Developing Distractors

Developing effective distractors (incorrect answer choices) is one of the hardest aspects of creating test items. An expert should be able to select the best answer easily, but a student with poor conceptual understanding should find at least one of the distractors attractive. Whenever possible, we based distractors on misconceptions uncovered during our student interviews [45]. Concretely doing so was not always possible in part because the interviews did not cover all of the scenarios ultimately created, so we also based distractors on general misconceptions (e.g., encryption solves everything). The difficulty is to develop enough distractors while satisfying the many constraints and objectives.

For simplicity, we decided that each test item would have exactly one best (but not necessarily ideal) answer. To simplify statistical analysis of our assessments [31], we decided that each test item would have the same number of alternatives. To reduce the likelihood of guessing correctly, and to reduce the required number of test items, we also decided that the number of alternatives would be exactly five. There is no compelling requirement to use five; other teams might choose a different number (e.g., 2–6).

Usually, it is fairly easy to think of two or three promising distractors. The main difficulty is coming up with the fourth. For this reason, test creators might choose to present four rather than five alternatives. Using only four alternatives (versus five) increases the likelihood of a correct guess; nevertheless, using four alternatives would be fine, provided there are enough test items to yield the desired statistical confidence in student scores.

As we do when drafting stems, we consider the "Vanderbilt" guidelines [2], which include the following: All alternatives should be plausible (none should be silly), and each distractor should represent some misconception. Each alternative should be as short as reasonably possible. The alternatives should be mutually exclusive (none should overlap). The alternatives should be relatively homogeneous (none should stand out as different, for example, in structure, format, grammar, or length). If all alternatives share a common word or phrase, that phrase should be moved to the stem.

Care should be taken to avoid leaking clues, both within a test item and between different test items. In particular, avoid leaking clues with strong diction, length, or any unusual difference among alternatives. Never use the alternatives "all of the above" or "none of the above;" these alternatives complicate statistical analysis and provide little insight into student understanding of concepts. If a negative word (e.g., "**NOT**") appears in a test item, it should be emphasized to minimize the chance of student misunderstandings. As noted in Sect. 3.6, typically stems should be worded in a positive way.

To develop distractors, we used the same interactive iterative process described in Sect. 3.6. We would begin with the correct alternative, which for our convenience only during test item development, we always listed as Alternative A. Sometimes we would develop five or six distractors, and later pick the four selected most frequently by students. To overcome issues (e.g., ambiguity, possible multiple best answers, or difficulty coming up with more distractors), we usually added more details to the scenario or stem. For example, to constrain the problem, we might clarify the assumptions or adversarial model.

Reflecting on the difficulty of conducting student interviews and brainstorming quality distractors, we investigated alternate ways to develop distractors [38,39]. One way is to have students from the targeted population answer stems *without* being offered any alternatives. By construction, popular incorrect answers are distractors that some subjects will find attractive. This method has the advantage of using a specific actual stem. For some test items, we did so using student responses from our student interviews. We could not do so for all test items because we created some of our stems after our interviews.

An even more intriguing variation is to collect such student responses through crowdsourcing (e.g., using Amazon Mechanical Turk [27]). We were able to do so easily and inexpensively overnight [38,39]. The main challenges are the inability to control the worker population adequately, and the high prevalence of cheaters (e.g., electronic bots deployed to collect worker fees, or human workers who do not expend a genuine effort to answer the stem). Nevertheless, even if the overwhelming majority of responses are gibberish, the process is successful if one can extract at least four attractive distractors. Regardless, the responses require grouping and refinement. Using crowdsourcing to generate distractors holds great promise and could be significantly improved with verifiable controls on the desired workers.

Continuing the switchbox example from Sects. 3.5 and 3.6, we explain how we drafted the distractors and how they evolved. Originally, when we had created the scenario, we had wanted the correct answer to be preventing data from being exfiltrated from the accounting LAN (Alternative D is a more specific instance of this idea). Because the system design does not prevent this action, we settled on the correct answer being preventing access to the accounting LAN. To make the correct answer less obvious, we worded it specifically about employees accessing the accounting LAN from home. Intentionally, we chose not to use a broader wording about people accessing the accounting LAN from the Internet, which subjects in our new crowdsourcing experiment (Sect. A) subsequently came up with and preferred when presented the open-ended stem without any alternatives.

What security goal is this design primarily intended to meet?
A Prevent employees from accessing accounting data from home.
B Prevent the accounting LAN from being infected by malware.
C Prevent employees from using wireless connections in Accounting.
D Prevent accounting data from being exfiltrated.
E Ensure that only authorized users can access the accounting network.

Initial Version

Choose the action that this design best prevents:
A Employees accessing the accounting server from home.
B Infecting the accounting LAN with malware.
C Computer A communicating with Computer B.
D Emailing accounting data.
E Accessing the accounting network without authorization.

Final Version

Fig. 3. Evolution of the alternatives for the CCA switchbox test item.

Figure 3 illustrates the evolution of five alternatives to this CCA stem. As discussed in Sect. 3.6, the initial version of this test item has undesirable ambiguity. A second issue is that that Alternative C is highly implausible because there is nothing in the scenario or stem that involves wireless connections. A third issue is that the word "prevent" appears in four alternatives.

The final version of the test item makes three improvements. First, we cast the stem and alternatives in a more concrete and less ambiguous fashion. Second, Alternative C appears more plausible. Third, we moved the word "prevent" into the stem. In each version, the best answer is Alternative A.

Reflecting on data from our new crowdsourcing experiment (Sect. A), the problem development committee met to revisit the switchbox example again. Given that significantly more respondents chose Alternative E over A, we wondered why and carefully reexamined the relative merits of these two competing alternatives. Although we still prefer Alternative A over E (mainly because E emphasizes *authorization*, while the switch deals only with physical *access*), we recognize that E has some merit, especially with regard to malicious activity originating from the Internet. To make Alternative A unarguably better than E, we reworded E more narrowly: "User of Computer B from accessing the accounting LAN without authorization."

We made this change to Alternative E, not per se because many subjects chose E over A, but because we recognized a minor issue with Alternative E. The experimental data brought this issue to our attention. It is common in concept inventories that, for some test items, many subjects will prefer one of the distractors over the correct alternative, and this outcome is fine.

3.8 Generating Educational Materials

Pleased with the student engagement stimulated by our scenarios and associated interview prompts, we realized that these scenarios can make effective case studies through which students can explore cybersecurity concepts. To this end, we published a paper presenting six of our favorite scenarios, together with our exemplary responses and selected misconceptions students revealed reasoning about these scenarios [42]. These scenarios provide an excellent way to learn core concepts in cybersecurity in thought-provoking complex practical contexts.

Although we have not yet done so, it would be possible to create additional learning activities inspired by these scenarios, including structured discussions, design and analysis challenges, and lab exercises.

Also, one could prepare a learning document that consisted of new MCQ test items together with detailed discussions of their answers. For such a document, it would be helpful to permit 0–5 correct answers for each test item. Doing so would reap two benefits: First, it would simplify creating test items because it would avoid the challenge of requiring exactly one best answer. Second, it would create an authentic learning moment to be able to discuss a variety of possible answers. In particular, typically there can be many possible ways to design, attack, and defend a system.

3.9 Performing Expert Reviews

We obtained feedback from cybersecurity experts on our draft assessments in three ways. First, we received informal feedback. Second, experts reviewed the CCI and CCA during our workshops and hackathons. Third, as part of our validation studies, we conducted more formal expert reviews [31]. These experts include cybersecurity educators and practitioners from government and private industry. We recruited the experts through email, web announcements, and at conferences.

Each expert took the CCI or CCA online (initially through SurveyMonkey, and later via PrairieLearn [48]). For each test item, they first selected their answer choice. Next, we revealed what we considered to be the correct answer. We then invited the expert to write comments, and the expert rated the test item on the following scale: accept, accept with minor revisions, accept with major revisions, reject. Following the online activity, we engaged a group of experts in a discussion about selected test items.

We found this feedback very useful. It was reassuring to learn that the experts agreed that the test items probe understanding of the targeted concepts and that they agreed with our answer choices. The experts helped us identify issues with some of the test items. Because resolving such issues can be very time consuming, we preferred to keep the discussion focused on identifying the issues. In days following an expert review, the problem development group refined any problematic test items.

Some experts disagreed with a few of our answer choices. Usually they changed their opinion after hearing our explanations. Sometimes the disagreement reflected an ambiguity or minor error in the test item, which we later resolved, often by providing more details in the scenario. Usually, when experts disagreed with our answer choice, it was because they made an extra assumption not stated in the scenario or stem. We instructed subjects not to make any unstated assumptions, but some common sense assumptions are often required. Experts, more so than students, struggled with the challenge of what to assume, possibly because of their extensive knowledge of possibly relevant factors. Dealing with this challenge is one of the special difficulties in creating cybersecurity concept inventories.

3.10 Recruiting Test Subjects

Recruiting large numbers of test subjects turned out to be much more challenging than we had expected. We are validating the CCI in 2018–2020 through small-scale pilot testing (100–200 subjects), cognitive interviews, expert review, and large-scale psychometric testing (1000 subjects). For the CCA, we are following a parallel plan in 2019–2021. Staggering the validations of the CCI and CCA helps balance our work and permits us to adapt lessons learned from the earlier years. It is difficult to recruit others to invest their scarce time helping advance our project.

For pilot testing of the CCI, we recruited 142 students from six schools. We also recruited 12 experts and carried out approximately seven cognitive interviews. We recruited these students by direct contact with educators we knew who were teaching introductory cybersecurity classes. The overwhelming majority of experts rated every item as measuring appropriate cybersecurity knowledge. Classical test theory [1,21,26] showed that the CCI is sufficiently reliable for measuring student understanding of cybersecurity concepts and that the CCI may be too difficult as a whole [31]. In response to these inputs, we revised the CCI and moved one of the harder test items to the CCA.

In fall 2019, we started to recruit 1000 students to take the CCI so that we can evaluate its quality using item response theory [1,21,26], which requires many more subjects than does classical test theory. Recruiting subjects proved difficult and by late December just under 200 students had enrolled. We hope to recruit another 800 subjects by late spring 2020. Because of the ongoing COVID-19 pandemic, all of these subjects will take the CCI online.

By far, the most effective recruitment method has been direct contact with educators (or their associates) we personally know who teach introductory cybersecurity classes. Sending email to people we do not personally know has been extremely ineffective, with a response rate of approximately 1%. Making personal contact at conferences usually resulted in a stated willingness to participate but without subsequent action. Following up on such promises sometimes increased our yield. We advertised through email, web, and conferences. Posting notices in newsletters (e.g., for CyberWatch), and making announcements at PI meetings seemed to have some positive effect. We targeted specific likely groups, including people connected with Scholarship for Service (SFS) programs and National Centers of Excellence in Cyber Defense Education (CAE). We also asked participating instructors for referrals to additional instructors. At conferences, passing out slips of paper with URLs and QR-codes did not work well. Keeping our recruitment pitch brief, limited, and simple helped.

For instructors who enroll in our validation study, some administer the CCI in class through our web-based system; others suggest it as an optional out-of-class activity. For the latter group, offering some type of incentive (e.g., extra credit) has been critical. It takes students approximately one hour to complete the CCI and two hours to take the CCA. Virtually no students took the CCI in classes where the instructor offered no incentive.

Timing has been another issue. Some colleges and universities offer an introductory cybersecurity course only once a year, and students are not ready to take the assessment until towards the end of the course. Furthermore, contacting students after they have completed the course has been mostly ineffective. It has improved our yield to ask an instructor when they will be able to administer the assessment, and then follow up when the time comes closer.

Recruiting 1000 subjects for the CCA might be even harder because the target population is smaller.

3.11 Organizing and Running Hackathons

In February 2018, we hosted a two-day "Hackathon" for 17 cybersecurity educators and professionals from across the nation to generate multiple-choice test items for the CCA, and to refine draft items for the CCI and CCA [43]. To collect additional expert feedback, we also held shorter workshops at the 2019 National Cyber summit and 2019 USENIX Security Symposium. We had to cancel another workshop planned for June 2020, to have been coordinated with the Colloquium for Information System Security (CISSE), due to the COVID-19 pandemic.

Our focused Hackathon engaged experts, generated useful inputs, and promoted awareness of the project. Each participant focused on one of the following tasks: 1) generating new scenarios and stems; 2) extending CCI items into CCA items, and generating new answer choices for new scenarios and stems from Task 1; and 3) reviewing and refining draft CCA test items. These tasks kept each team fully engaged throughout the Hackathon. Each participant chose what team to join, based in part on their skill sets. The event took place at an off-campus conference center, two days before the ACM Special Interest Group on Computer Science Education (SIGSCE) conference in Baltimore, Maryland. Thirteen experts came from universities and two each came from industry and government, respectively. Participants took the CCI at the beginning of the first day and the CCA at the beginning of the second day. We paid participant costs from our NSF grant.

The most difficult challenge in organizing our Hackathon was obtaining commitments of participants to attend. We had originally planned to hold the event in October 2017, but rescheduled due to low interest. Holding the event immediately before a major computer science education conference helped. As for recruiting instructors in our pilot studies, direct one-on-one person contact was our most effective communication method.

With the hopes of more easily attracting participants, for the shorter workshops we tried a slightly different strategy. We coordinated with conference organizers to hold a four-hour workshop at the conference site, typically immediately before the main conference began. Although recruitment remained a challenge and the shorter workshop provided fewer inputs, this strategy mostly worked better. To our surprise, offering food at these shorter workshops seemed to make no difference.

Initially, we had thought that offering to pay participant expenses (airfare, hotel, per-diem) would help motivate participants to attend. This strategy, however, encountered two issues. First, many people already had funding and they could not absorb additional reimbursements. Second, reimbursements require significant paperwork and administrative procedures, which are inconvenient for participants and organizers. When possible, it is simpler and more effective instead to pay honoraria.

3.12 Building Community Acceptance

Many concept inventories remain largely unused, poorly accepted, and poorly adopted. Members of the subject community do not recognize the importance of the concepts they assess or the effectiveness of the assessments. To avoid this fate, great care must be taken to build acceptance throughout the creation and validation process.

The CATS Project is pursuing three strategies to build widespread acceptance. First, we are following a well-accepted scientific methodology for creating and validating the CCI and CCA. Second, throughout the project we are engaging experts, including in our Delphi processes that identified core concepts, in our workshops and hackathons that develop and refine test items, and in our experts reviews of draft assessments. Third, we are presenting our work at conferences to seek feedback and promote awareness of our assessment tools.

It remains to be seen how widely the CCI and CCA will be used. One special characteristic that may help is the driving recognition of the strong need to prepare more cybersecurity practitioners. It would probably further advance our assessments to carry out an even more active program of presenting our work to a wide range of cybersecurity educators. We hope that many researchers will use the CCI and CCA to identify and measure effective approaches to teaching and learning cybersecurity.

3.13 Forming a Team and Nurturing Collaboration

With experts in cybersecurity, engineering education, and educational assessment, our team covers all of the key areas needed to accomplish our goals. In particular, Herman brings prior experience creating, validating, and using concept inventories. A postdoc, several graduate students, an undergraduate student, and a high school student also add useful diversity, talent, and labor. For example, one PhD student brings practical cybersecurity experience from NSA and private industry. This interdisciplinary, inter-institution collaboration including a variety of perspectives has been crucial to our success.

The entire team meets every Thursday on Skype for up to an hour, whenever there is work to be done; we skip meetings if there is nothing to discuss. These meetings help us make strategic decisions and keep us focused on what needs to be done. In addition, subgroups—such as the problem development group—meet weekly as needed. Oliva assumed primary responsibility for handling all IRB approvals and conducting student interviews. Sherman and Herman took primary responsibility for writing all grant proposals and grant reports.

When developing a test item, the problem development group might spend time discussing one or more of the following steps: high-level brainstorming, creating a scenario, drafting a stem, developing distractors, refining test items, and recording meta-data. A typical session lasted about two hours or slightly longer. After initially meeting in person at UMBC, we ultimately preferred to meet remotely while simultaneously editing a Google Doc, which was more efficient and accommodated new team member Peterson from Duluth, MN. We felt productive if we could make substantial progress on at least one test item per meeting.

When writing a paper, one team member would serve as the primary writer/editor. We delegated tasks such as writing particular sections, preparing figures, programming, and dealing with references.

Our team functions smoothly and has avoided any major fracas. Team members have mutual respect for each other and value each person's unique contributions. We resolve conflict in a civil and constructive way. We include everyone in deliberations about all important matters, encouraging frank critical discussions regardless of rank. This style of decision making helps us make wise choices.

3.14 Adopting Tools

We used a variety of tools to support various aspects of the project, including tools for conferencing (Skype), document sharing (Google Drive, Github, Overleaf), simultaneous editing (Google Docs, Overleaf), surveys (SurveyMonkey), document preparation (LaTeX with Overleaf, Microsoft Word), data management (Excel), and delivery and analysis of tests (PrairieLearn). We are not familiar with any useful dedicated tool for supporting the development of MCQs. To avoid the trouble of making and maintaining our own tool, we used a collection of existing tools. Over time, as we worked on different tasks and experimented with a variety of approaches for interacting with each other, our operational methods and choice of supporting tools evolved. Three critical tasks presented especially important challenges: conferencing to create and refine test items, delivery and analysis of tests, and management of test items throughout their life cycle.

Initially, we brainstormed test items at UMBC sitting around a conference table, while one team member wrote notes into an electronic document that we projected onto a screen. Eventually we found this method of operations suboptimal for several reasons. First, it was difficult to schedule a time when everyone in the problem-development group could attend. Second, it was inefficient for people to drive to UMBC for such meetings, especially because one member lived in DC and another moved to Richmond. Third, having only one person edit was inefficient, especially when making simple edits. Fourth, as often happens by human nature, we tended to waste more time when meeting in person than meeting online. Work proceeded more smoothly and efficiently when we switched to online conferencing, during which each participant simultaneously viewed and edited a Google Doc. Participants could see displays more clearly when looking at their own monitor than looking at a conference room projection. For groups that meet in person, we recommend that each participant bring a laptop to enable simultaneous editing.

When we started our validation studies, we needed a way for students and experts to take our assessments and for experts to comment on them. After experimenting briefly with SurveyMonkey, we chose PrairieLearn [48], a tool developed at the University of Illinois, and already used by Herman. This tool allows subjects to take tests online and provides statistical support for their analysis. It also supports a number of useful test delivery features, including randomizing answer choices. PrairieLearn, however, does not support the life cycle of test items. Also, in PrairieLearn it is inconvenient to enter test items with mathematical expressions (we awkwardly did so using HTML).

The CATS Project could have benefited greatly from an integrated tool to support the entire life cycle of test items, maintaining the authoritative version of each test item and associated meta-data (comments, difficulty, topics, concepts, and validation statistics), and avoiding any need to copy or translate test items from one system to another (which can introduce errors). Such a system should also support simultaneous editing, and it should display test items as they will appear to students. While developing more than 60 test items over several years, it was essential to save detailed notes about each test item in an orderly fashion.

Though PrairieLearn has many strengths, PrairieLearn falls short: with test items written in HTML, it was difficult to edit the source simultaneously. We tried using the Google-recommended HTML editor Edity but found it cumbersome, due in part to synchronization difficulties across multiple editors. We ended up making most edits in a Google Doc and then manually translating the results into PrairieLearn. Curiously, PrairieLearn provides no automatic way to generate a PDF file of a test, so we wrote a script to do so semi-automatically, by converting the PrairieLearn HTML to LaTeX using pandoc. Although we can define assessments in PrairieLearn, the system provides no high-level support for helping us to decide which test items (from a larger bank of items) should be included to achieve our desired goal of having five test items of various difficulty levels from each of the five targeted concepts.

Initially we collaborated writing documents using Github but eventually found Overleaf much easier to use. We prepared most of our publications using LaTeX. A few times we experimented with Microsoft Word, each time regretting our choice (due in part to poor support of mathematics, simultaneous editing, and fine-grain document control). We found Overleaf very convenient to use because it supports simultaneous editing and it provides a uniform compilation environment.

Because long email threads can be confusing, we are considering deploying Slack to manage channels of text messages that persist beyond the live exchanges. Such channels would be useful for many tasks of the project, including writing papers.

3.15 Obtaining and Using Funding

Three grants (each collaborative between UMBC and Illinois) have directly funded the CATS project: two one-year CAE-R grants from NSA, and one three-year SFS capacity grant from NSF. In addition, UMBC's main SFS grant provided additional support. Funds have supported a post-doc, graduate RAs, faculty summer support, travel, and workshops. This funding has been helpful in promoting collaborations, including between our two institutions, within each institution, and more broadly through workshops. Especially initially, the prospects of possible funding helped Sherman and Herman forge a new partnership.

Some of the factors that likely contributed to our successful funding include the following: We have a collaborating team that covers the needed areas of expertise. We submitted convincing detailed research plans (for example, we secured agreements with most of the Delphi experts by proposal time). At each step, we presented preliminary accomplishments from the previous steps. NSF recognized the value of a concept inventory that could be used as a scientific instrument for measuring the effectiveness of various approaches to cybersecurity education (see Sect. 3.1).

The CATS Project has produced much more than MCQs, and it has created more MCQs than the 50 on the CCI and CCA. For example, the CATS Project has published eight research papers, including on core concepts of cybersecurity [33], student misconceptions about cybersecurity [45], using crowdsourcing to generate distractors [38,39], and case studies for teaching cybersecurity [42]. Keeping in mind that most of our funding has supported research activities beyond creating MCQs, it is nevertheless interesting to try to approximate the cost of creating validated test items. Dividing the project's total combined funding by the 50 MCQs on the CCI and CCA yields a cost of $21,756 per test item. There is commercial potential for creating cybersecurity assessment tools, but any financially successful company would have to develop efficient processes and amortize its expenses over a large number of test takers.

4 Discussion

Working on the CATS Project we have learned a lot about creating and validating MCQ concept inventories. We now discuss some of our most notable takeaways from our evolving relationship with MCQs. In particular, we discuss the difficulty of creating effective MCQs, the special challenges of cybersecurity MCQs, the value of using scenarios in test items, the advantages and limitations of MCQs, and some general advice for creating test items.

Creating MCQs that effectively reveal mastery of core concepts is difficult and time consuming. Although we have become more efficient, it still takes us many hours to conceive of, draft, and refine a test item. Great care is needed to ensure that the test item is clear to all, there is exactly one best answer from among five plausible alternatives, students who understand the underlying concept will be able to select the correct answer, and students who do not understand the underlying concept will find some of the distractors appealing. Test items should not be informational or measures of intelligence. Care must be taken to minimize the chance that some students might become confused for unintended reasons that do not necessarily relate to conceptual understanding, such as about a particular detail or word choice.

Cybersecurity presents some special difficulties for creating and answering MCQs. First, details often matter greatly. In cybersecurity, issues are frequently subtle, and the best answer often hinges on the details and adversarial model. It is important to provide enough details (but not too many), and doing so is challenging within the constraints of a MCQ. Second, it is often difficult to ensure that there is exactly one best answer. For example, typically there are many potential designs and vulnerabilities. An "ideal" answer, if it existed, might appear too obvious. To deal with this difficulty, we usually add details to the scenario and stem; we emphasize that the task is to select the best alternative (which might not be a perfect answer) from the available choices; and we sometimes frame the question in terms of comparing alternatives (e.g., best or worst). Third, it is problematic to create MCQs questions based on attractive open-ended questions, such as, "Discuss potential vulnerabilities of this system." Often it can be much more challenging to think of a vulnerability than to recognize it when stated as an alternative. Listing a clever vulnerability or attack as an answer choice can spoil an otherwise beautiful question. Again, to deal with this difficulty, it can be helpful to ask students to compare alternatives. These difficulties are common to other domains (e.g., engineering design), and they are strongly present in cybersecurity.

We found it helpful to create test items that comprise a scenario, stem, and alternatives. Each scenario describes an engaging situation in sufficient detail to motivate a meaningful cybersecurity challenge. For the CCA, to add technical substance, we typically included some artifact, such as a protocol, system design, log file, or source code. Our strategy was to put most of the details into the scenario—rather than into the stem—so that the stem could be as short and straightforward as reasonably possible. This strategy worked well. We created an initial set of twelve scenarios for our talk-aloud interviews to uncover student misconceptions [45]. The scenarios provide useful building blocks for other educational activities, including exercises and case studies [42]. In creating scenarios, it is necessary to balance depth, breadth, richness, and length.

Originally, we had expected to be able easily to create several different stems for each scenario, which should reduce the time needed to complete the assessment and reduce the cognitive load required to read and process details. For example, in the *Scholastic Assessment Test (SAT)*, several test items share a common reading passage. Although several of our test items share a common scenario, unexpectedly we found it problematic to do so for many scenarios. The reason is that, for each stem, we felt the need to add many details to the scenario to ensure that there would be exactly one best answer, and such details were often specific to the stem. After customizing a scenario in this way, the scenario often lost the generality needed to work for other stems. Whenever we do share a scenario, for clarity, we repeat it verbatim.

In keeping with the predominant format of CIs, we chose to use MCQs because they are relatively fast and easy to administer and grade; there is a well established theory for validating them and interpreting their results [34]; it is possible to create effective MCQs; and they are relatively easier to create than are some alternatives (e.g., computer simulations). Albeit more complex and expensive, alternatives to MCQs may offer ways to overcome some of the limitations of MCQs. These alternatives include open-ended design and analysis tasks, hands-on challenges, games, and interactive simulations. We leave as open problems to create and validate cybersecurity assessment tools based on alternatives to MCQs.

To generate effective MCQ test items, we found it useful to work in a diverse collaborating team, to draw from our life experiences, to engage in an iterative development process, and to be familiar with the principles and craft of drafting questions and the special challenges of writing multiple-choice questions about security scenarios. It is helpful to become familiar with a variety of question types (e.g., identifying the best or worst), as well as question themes (e.g., design, attack, or defend). When faced with a difficult challenge, such as reducing ambiguity or ensuring a single best answer, we found the most useful strategy to be adding details to the scenario and stem, to clarify assumptions and the adversarial model. Introducing artifacts is an engaging way to deepen the required technical knowledge. To promote consistency, we maintained a style guide for notation, format, and spelling.

5 Conclusion

When we started the CATS Project six years ago, we would have loved to have been able to read a paper documenting experiences and lessons learned from creating and validating a concept inventory. The absence of such a paper motivated us to write this one. As we addressed each key step of the process creating two cybersecurity concept inventories, we selected an approach that seemed appropriate for our situation, adjusting our approach as needed.

One of the most important factors contributing to our success is collaboration. Our diverse team includes experts in cybersecurity, systems, engineering education, and educational assessment. When creating and refining test items, it is very helpful to have inputs from a variety of perspectives. Often in security, a small observation makes a significant difference. Openness and a willingness to be self-critical helps our team stay objectively focused on the project goals and tasks at hand.

When we started the project, many of us had a very poor opinion of MCQs, based primarily on the weak examples we had seen throughout our lives. Many MCQs are informational, thoughtless, and flawed in ways that can permit answering questions correctly without knowledge of the subject.[3] Over time, we came to realize that it is possible to create excellent conceptual MCQs, though doing so is difficult and time consuming.

We believe strongly that the core of cybersecurity is *adversarial thinking*—managing information and trust in an adversarial cyber world, in which there are malicious adversaries who aim to carry out their nefarious goals and defeat the objectives of others. Appropriately, our assessment tools focus on five core concepts of adversarial thinking, identified in our Delphi studies.

Our work on the CATS Project continues as we complete the validation and refinement of the CCI and CCA with expert review, cognitive interviews, small-scale pilot testing, and large-scale psychometric testing. We invite and welcome you to participate in these steps.[4] Following these validations, we plan to apply these assessment tools to measure the effectiveness of various approaches to teaching and learning cybersecurity.

We hope that others may benefit from our experiences with the CATS Project.

Acknowledgments. We thank the many people who contributed to the CATS project as Delphi experts, interview subjects, Hackathon participants, expert reviewers, student subjects, and former team members, including Michael Neary, Spencer Offenberger, Geet Parekh, Konstantinos Patsourakos, Dhananjay Phatak, and Julia Thompson. Support for this research was provided in part by the U.S. Department of Defense under CAE-R grants H98230-15-1-0294, H98230-15-1-0273, H98230-17-1-0349, H98230-17-1-0347; and by the National Science Foundation under UMBC SFS grants DGE-1241576, 1753681, and SFS Capacity Grants DGE-1819521, 1820531.

A A Crowdsourcing Experiment

To investigate how subjects respond to the initial and final versions of the CCA switchbox question (see Sects. 3.5–3.7), we polled 100 workers on Amazon Mechanical Turk (AMT) [27]. We also explored the strategy of generating distractors through crowdsourcing, continuing a previous research idea of ours [38,39]. Specifically, we sought responses from another 200 workers to the initial and final stems without providing any alternatives.

On March 29–30, 2020, at separate times, we posted four separate tasks on AMT. Tasks 1 and 2 presented the CCA switchbox test item with alternatives, for the initial and final versions of the test item, respectively. Tasks 3 and 4 presented the CCA switchbox stem with *no* alternatives, for the initial and final versions of the stem, respectively.

[3] As an experiment, select answers to your favorite cybersecurity certification exam looking only at the answer choices and not at the question stems. If you can score significantly better than random guessing, the exam is defective.

[4] Contact Alan Sherman (sherman@umbc.edu).

For Tasks 1–2, we solicited 50 workers each, and for Tasks 3–4, we solicited 100 workers each. For each task we sought human workers who graduated from college with a major in computer science or related field. We offered a reward of \$0.25 per completed valid response and set a time limit of 10 min per task.

Expecting computer bots and human cheaters (people who do not make a genuine effort to answer the question), we included two control questions (see Fig. 4), in addition to the main test item. Because we expected many humans to lie about their college major, we constructed the first control question to detect subjects who were not human or who knew nothing about computer science. All computer science majors should be very familiar with the binary number system. We expect that most bots will be unable to perform the visual, linguistic, and cognitive processing required to answer Question 1. Because $11 + 62 = 73 = 64 + 8 + 1$, the answer to Question 1 is 1001001.

We received a total of 547 responses, of which we deemed 425(78%) valid. As often happens with AMT, we received *more* responses than we had solicited, because some workers responded directly without payment to our SurveyMonkey form, bypassing AMT. More specifically, we received 40, 108, 194, 205 responses for Tasks 1–4, respectively, for which 40(100%), 108(100%), 120(62%), 157(77%) were valid, respectively. In this filtering, we considered a response valid if and only if it was non-blank and appeared to answer the question in a meaningful way. We excluded responses that did not pertain to the subject matter (e.g., song lyrics), or that did not appear to reflect genuine effort (e.g., "OPEN ENDED RESPONSE").

Only ten (3%) of the valid responses included a correct answer to Question 1. Only 27(7%) of the workers with valid responses stated that they majored in computer science or related field, of whom only two (7%) answered Question 1 correctly. Thus, the workers who responded to our tasks are very different from the intended population for the CCA.

Originally, we had planned to deem a response valid if and only if it included answers to all three questions, with correct answers to each of the two control questions. Instead, we continued to analyze our data adopting the extremely lenient definition described above, understanding that the results would not be relevant to the CCA.

Question 1: What is eleven plus 62? Express your answer in binary.

Question 2: In what major did you graduate from college?
A English
B History
C Chemistry
D Computer science or related field
E Business
F I did not graduate from college

Fig. 4. Control questions for the crowdsourcing experiment on AMT. Question 1 aims to exclude bots and subjects who do not know any computer science.

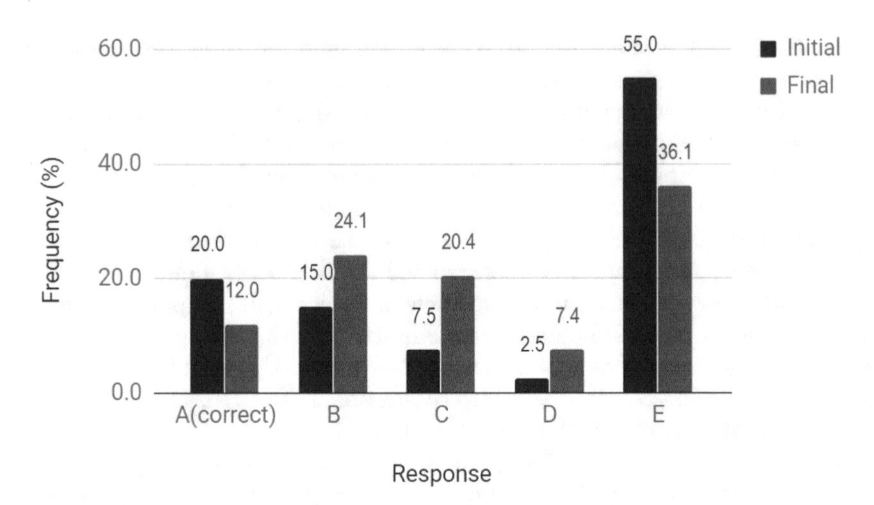

Fig. 5. Histogram of responses to the CCA switchbox test item, from AMT workers, for the initial and final versions of the test item. There were 148 valid responses total, 40 for the initial version, and 108 for the final version.

We processed results from Tasks 3–4 as follows, separately for each task. First, we discarded the invalid responses. Second, we identified which valid responses matched existing alternatives. Third, we grouped the remaining (non-matching) valid new responses into equivalency classes. Fourth, we refined a canonical response for each of these equivalency classes (see Fig. 8). The most time-consuming steps were filtering the responses for validity and refining the canonical responses.

Especially considering that the AMT worker population differs from the target CCA audience, it is important to recognize that the best distractors for the CCA are not necessarily the most popular responses from AMT workers. For this reason, we examined all responses. Figure 8 includes two examples, $t*$ for initial version (0.7%) and $t*$ for final version (1.2%), of interesting *unpopular* responses. Also, the alternatives should satisfy various constraints, including being non-overlapping (see Sect. 3.7), so one should not simply automatically choose the four most popular distractors.

Figures 5, 6, 7 and 8 summarize our findings. Figure 5 does not reveal any dramatic change in responses between the initial and final versions of the test item. It is striking how strongly the workers prefer Distractor E over the correct choice A. Perhaps the workers, who do not know network security, find Distractor E the most understandable and logical choice. Also, its broad wording may be appealing. In Sect. 3.7, we explain how we reworded Distractor E to ensure that Alternative A is now unarguably better than E.

Figure 6 shows the popularity of worker-generated responses to the stem, when workers were *not* given any alternatives. After making Fig. 6, we realized it contains a mistake: responses $t2$ (initial version) and $t1$ (final version) are *correct* answers, which should have been matched and grouped with Alternative A. We

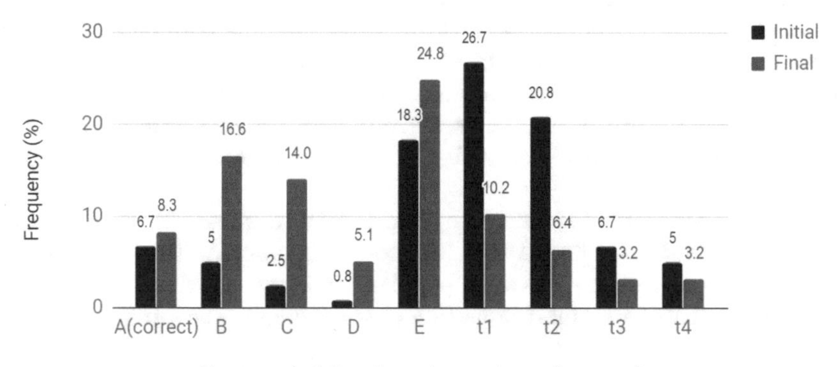

Fig. 6. Histogram for equivalency classes of selected open-ended responses generated from the CCA switchbox stem. These responses are from AMT workers, for the initial and final versions of the stem, presented without any alternatives. There were 120 valid responses for the initial version, and 157 for the final version. A–E are the original alternatives, and $t1$–$t4$ are the four most frequent new responses generated by the workers. These responses include two alternate phrasings of the correct answer: $t2$ for the initial version, and $t1$ for the final version. Percents are with respect to all valid responses and hence do not add up to 100%.

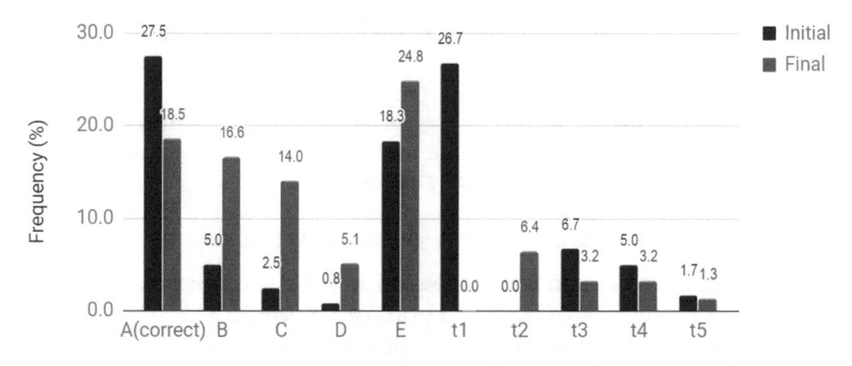

Fig. 7. Histogram for equivalency classes of selected open-ended responses generated from the CCA switchbox stem. These responses are from AMT workers, for the initial and final versions of the stem, presented without any alternatives. There were 120 valid responses for the initial version, and 157 for the final version. A–E are the original alternatives, and $t1$–$t5$ are the four most frequent new distractors generated by the workers. The alternate correct responses $t2$ (initial version) and $t1$ (final version) are grouped with Alternative A. Percents are with respect to all valid responses and hence do not add up to 100%.

nevertheless include this figure because it is informative. Especially for the initial version of the stem, it is notable how more popular the two worker-generated distractors $t2$ (initial version) and $t1$ (final version) are than our distractors. This finding is not unexpected, because, in the final version, we had intentionally chosen Alternative A over $t1$, to make the correct answer less obvious (see Sect. 3.7). These data support our belief that subjects would find $t1$ more popular than Alternative A.

Figure 7 is the corrected version of Fig. 6, with responses $t2$ (initial version) and $t1$ (final version) grouped with the correct answer A. Although new distractor $t1$ (initial version) was very popular, its broad nebulous form makes it unlikely to contribute useful information about student conceptual understand-

What security goal is this design primarily intended to meet?

A Prevent employees from accessing accounting data from home. (correct)
B Prevent the accounting LAN from being infected by malware.
C Prevent employees from using wireless connections in Accounting.
D Prevent accounting data from being exfiltrated.
E Ensure that only authorized users can access the accounting network.

t1 Ensure all segments of the network remain independent.
t2 Keep the accounting server separated from the Internet. (also correct)
t3 Prevent malware from affecting the network.
t4 Prevent the email server from being compromised.
t5 Ensure the security of computer B.
t* Protect accounting server from DDoS attacks.

Initial Version

Choose the action that this design best prevents:

A Employees accessing the accounting server from home. (correct)
B Infecting the accounting LAN with malware.
C Computer A communicating with Computer B.
D Emailing accounting data.
E Accessing the accounting network without authorization.

t1 Accounting server from being accessed through the Internet. (also correct)
t2 A computer from accessing both Local Area Networks.
t3 Accessing the LAN via the Network Interface Card.
t4 Email server from being compromised.
t5 Computer C from accessing the Internet.
t* Unauthorized access from computer A to server.

Final Version

Fig. 8. Existing alternatives (A–E), and refinements of the five most frequent new responses ($t1$–$t5$) generated for the CCA switchbox stem, from AMT workers, for the initial and final versions of the stem. There were 120 valid responses for the initial version, and 157 for the final version. Distractors $t*$ for initial version (0.7%) and $t*$ for final version (1.2%) are examples of interesting unpopular responses. Percents are with respect to all valid responses.

ing. For the first version of the stem, we identified seven equivalency classes of new distractors (excluding the new alternate correct answers); for the final version we identified 12.

Because the population of these workers differs greatly from the intended audience for the CCA, these results should not be used to make any inferences about the switchbox test item for the CCA's intended audience. Nevertheless, our experiment illustrates some notable facts about crowdsourcing with AMT. (1) Collecting data is fast and inexpensive: we collected all of our responses within 24 hours paying a total of less than $40 in worker fees (we did not pay for any invalid responses). (2) We had virtually no control over the selection of workers, and almost all of them seemed ill-suited for the task (did not answer Question 1 correctly). Nevertheless, several of the open-ended responses reflected a thoughtful understanding of the scenario. (3) Tasks 3–4 did produce distractors (new and old) of note for the worker population, thereby illustrating the potential of using crowdsourcing to generate distractors, if the selected population could be adequately controlled. (4) Even when the worker population differs from the desired target population, their responses can be useful if they inspire test developers to improve test items and to think of effective distractors.

Despite our disappointment with workers answering Question 1 incorrectly, the experience helped us refine the switchbox test item. Reflecting on the data, the problem development team met and made some improvements to the scenario and distractors (see discussions near the ends of Sects. 3.5 and 3.7). Even though the AMT workers represent a different population than our CCA target audience, their responses helped direct our attention to potential ways to improve the test item.

We strongly believe in the potential of using crowdsourcing to help generate distractors and improve test items. Being able to verify the credentials of workers assuredly (e.g., cryptographically) would greatly enhance the value of AMT.

References

1. Bechger, T.M., Maris, G., Verstralen, H., Beguin, A.A.: Using classical test theory in combination with item response theory. Appl. Psychol. Meas. **27**, 319–334 (2003)
2. Brame, C.J.: Writing good multiple choice test questions 2019. https://cft.vanderbilt.edu/guides-sub-pages/writing-good-multiple-choice-test-questions/. Accessed 19 Jan 2019
3. Brown, B.: Delphi process: a methodology used for the elicitation of opinions of experts. CA, USA, Rand Corporation, Santa Monica, September 1968
4. U.S. Department of Labor Bureau of Labor Statistics. Information Security Analysts. *Occupational Outlook Handbook*, September 2019
5. George Washington University Arlington Center. Cybersecurity education workshop, February 2014. https://research.gwu.edu/sites/g/files/zaxdzs2176/f/downloads/CEW_FinalReport_040714.pdf
6. Chi, M.T.H.: Methods to assess the representations of experts' and novices' knowledge. In: Cambridge Handbook of Expertise and Expert Performance, pp. 167–184 (2006)

7. Ciampa, M.: CompTIA Security+ Guide to Network Security Fundamentals, Loose-Leaf Version, 6th edn. Course Technology Press, Boston (2017)
8. CompTIA. CASP (CAS-003) certification study guide: CompTIA IT certifications
9. International Information System Security Certification Consortium: Certified information systems security professional. https://www.isc2.org/cissp/default.aspx. Accessed 14 Mar 2017
10. International Information Systems Security Certification Consortium. GIAC certifications: The highest standard in cyber security certifications. https://www.giac.org/
11. Cybersecurity and Infrastructure Security Agency. The National Initiative for Cybersecurity Careers & Studies. https://niccs.us-cert.gov/featured-stories/take-cybersecurity-certification-prep-course
12. Epstein, J.: The calculus concept inventory: measurement of the effect of teaching methodology in mathematics. Not. ACM **60**(8), 1018–1025 (2013)
13. Ericsson, K.A., Simon, H.A.: Protocol Analysis. MIT Press, Cambridge (1993)
14. Evans, D.L., Gray, G.L., Krause, S., Martin, J., Midkiff, C., Notaros, B.M., Pavelich, M., Rancour, D., Reed, T., Steif, P., Streveler, R., Wage, K.: Progress on concept inventory assessment tools. In 33rd Annual Frontiers in Education, 2003. FIE 2003, pp. T4G– 1, December 2003
15. CSEC2017 Joint Task Force. Cybersecurity curricula 2017. Technical report, CSEC2017 Joint task force, December 2017
16. Gibson, D., Anand, V., Dehlinger, J., Dierbach, C., Emmersen, T., Phillips, A.: Accredited undergraduate cybersecurity degrees: four approaches. Computer **52**(3), 38–47 (2019)
17. Glaser, B.G., Strauss, A.L., Strutzel, E.: The discovery of grounded theory: strategies for qualitative research. Nurs. Res. **17**(4), 364 (1968)
18. Goldman, K., Gross, P., Heeren, C., Herman, G.L., Kaczmarczy, L., Loui, M.C., Zilles, C.: Setting the scope of concept inventories for introductory computing subject. ACM Trans. Comput. Educ. **10**(2), 1–29 (2010)
19. Hake, R.: Interactive-engagement versus traditional methods: a six-thousand-student survey of mechanics test data for introductory physics courses. Am. J. Phys. **66**(1), 64–74 (1998)
20. Hake, R.: Lessons from the physics-education reform effort. Conserv. Ecol. **5**, 07 (2001)
21. Hambleton, R.K., Jones, R.J.: Comparison of classical test theory and item response theory and their applications to test development. Educ. Meas. Issues Pract. **12**, 253–262 (1993)
22. Herman, G.L., Loui, M.C., Zilles, C.: Creating the digital logic concept inventory. In: Proceedings of the 41st ACM Technical Symposium on Computer Science Education, pp. 102–106, January 2010
23. Hestenes, D., Wells, M., Swackhamer, G.: Force concept inventory. Phys. Teach. **30**, 141–166 (1992)
24. Jindeel, M.: I just don't get it: common security misconceptions. Master's thesis, University of Minnesota, June 2019
25. Association for computing machinery (ACM) joint task force on computing curricula and IEEE Computer Society. Computer Science Curricula: Curriculum Guidelines for Undergraduate Degree Programs in Computer Science. Association for Computing Machinery, New York (2013)
26. Jorion, N., Gane, B., James, K., Schroeder, L., DiBello, L., Pellegrino, J.: An analytic framework for evaluating the validity of concept inventory claims. J. Eng. Educ. **104**(4), 454–496 (2015)

27. Kittur, A., Chi, E.H., Suh, B.: Crowdsourcing user studies with Mechanical Turk. In: Proceedings of the SIGCHI Conference on Human Factors in Computing Systems, pp. 453–456, April 2008
28. Litzinger, T.A., Van Meter, P., Firetto, C.M., Passmore, L.J., Masters, C.B., Turns, S.R., Gray, G.L., Costanzo, F., Zappe, S.E.: A cognitive study of problem solving in statics. J. Eng. Educ. **99**(4), 337–353 (2010)
29. Mestre, J.P., Dufresne, R.J., Gerace, W.J., Hardiman, P.T., Touger, J.S.: Promoting skilled problem-solving behavior among beginning physics students. J. Res. Sci. Teach. **30**(3), 303–317 (1993)
30. NIST. NICE framework. http://csrc.nist.gov/nice/framework/. Accessed 8 Oct 2016
31. Offenberger, S., Herman, G.L., Peterson, P., Sherman, A.T., Golaszewski, E., Scheponik, T., Oliva, L.: Initial validation of the cybersecurity concept inventory: pilot testing and expert review. In: Proceedings of Frontiers in Education Conference, October 2019
32. Olds, B.M., Moskal, B.M., Miller, R.L.: Assessment in engineering education: evolution, approaches and future collaborations. J. Eng. Educ. **94**(1), 13–25 (2005)
33. Parekh, G., DeLatte, D., Herman, G.L., Oliva, L., Phatak, D., Scheponik, T., Sherman, A.T.: Identifying core concepts of cybersecurity: results of two Delphi processes. IEEE Trans. Educ. **61**(11), 11–20 (2016)
34. DiBello, L.V., James, K., Jorion, N. and Schroeder, L., Pellegrino, J.W.: Concept inventories as aids for instruction: a validity framework with examples of application. In: Proceedings of Research in Engineering Education Symposium, Madrid, Spain (2011)
35. Pellegrino, J., Chudowsky, N., Glaser, R.: Knowing What Students Know: The Science and Design of Educational Assessment. National Academy Press, Washington DC (2001)
36. Peterson, P.A.H, Jindeel, M., Straumann, A., Smith, J., Pederson, A., Geraci, B., Nowaczek, J., Powers, C., Kuutti, K.: The security misconceptions project, October 2019. https://secmisco.blogspot.com/
37. Porter, L., Zingaro, D., Liao, S.N., Taylor, C., Webb, K.C., Lee, C., Clancy, M.: BDSI: a validated concept inventory for basic data structures. In: Proceedings of the 2019 ACM Conference on International Computing Education Research, ICER 2019, pp. 111–119. Association for Computing Machinery, New York (2019)
38. Scheponik, T., Golaszewski, E., Herman, G., Offenberger, S., Oliva, L., Peterson, P.A.H., Sherman, A.T.: Investigating crowdsourcing to generate distractors for multiple-choice assessments (2019). https://arxiv.org/pdf/1909.04230.pdf
39. Scheponik, T., Golaszewski, E., Herman, G., Offenberger, S., Oliva, L., Peterson, P.A. and Sherman, A.T.: Investigating crowdsourcing to generate distractors for multiple-choice assessments. In: Choo, K.K.R., Morris, T.H., Peterson, G.L. (eds.) National Cyber Summit (NCS) Research Track, pp. 185–201. Springer, Cham (2020)
40. Scheponik, T., Sherman, A.T., DeLatte, D., Phatak, D., Oliva, L., Thompson, J. and Herman, G.L.: How students reason about cybersecurity concepts. In: IEEE Frontiers in Education Conference (FIE), pp. 1–5, October 2016
41. Offensive Security. Penetration Testing with Kali Linux (PWK). https://www.offensive-security.com/pwk-oscp/
42. Sherman, A.T., DeLatte, D., Herman, G.L., Neary, M., Oliva, L., Phatak, D., Scheponik, T., Thompson, J.: Cybersecurity: exploring core concepts through six scenarios. Cryptologia **42**(4), 337–377 (2018)

43. Sherman, A.T., Oliva, L., Golaszewski, E., Phatak, D., Scheponik, T., Herman, G.L., Choi, D.S., Offenberger, S.E., Peterson, P., Dykstra, J., Bard, G.V., Chattopadhyay, A., Sharevski, F., Verma, R., Vrecenar, R.: The CATS hackathon: creating and refining test items for cybersecurity concept inventories. IEEE Secur. Priv. **17**(6), 77–83 (2019)
44. Sherman, A.T., Oliva, L., DeLatte, D., Golaszewski, E., Neary, M., Patsourakos, K., Phatak, D., Scheponik, T., Herman, G.L. and Thompson, J.: Creating a cybersecurity concept inventory: a status report on the CATS Project. In: 2017 National Cyber Summit, June 2017
45. Thompson, J.D., Herman, G.L., Scheponik, T., Oliva, L., Sherman, A.T., Golaszewski, E., Phatak, D., Patsourakos, K.: Student misconceptions about cybersecurity concepts: analysis of think-aloud interviews. J. Cybersecurity Educ. Res. Pract. **2018**(1), 5 (2018)
46. Walker, M.: CEH Certified Ethical Hacker All-in-One Exam Guide, 1st edn. McGraw-Hill Osborne Media, USA (2011)
47. Wallace, C.S., Bailey, J.M.: Do concept inventories actually measure anything? Astron. Educ. Rev. **9**(1), 010116 (2010)
48. West, M., Herman, G.L. and Zilles, C.: PrairieLearn: mastery-based online problem solving with adaptive scoring and recommendations driven by machine learning. In: 2015 ASEE Annual Conference and Exposition, ASEE Conferences, Seattle, Washington, June 2015

Cyber Security Education

A Video-Based Cybersecurity Modular Lecture Series for Community College Students

Anton Dahbura(✉) and Joseph Carrigan

Johns Hopkins University, Baltimore, MD 21218, USA
{AntonDahbura,joseph.carrigan}@jhu.edu

Abstract. The shortage of people who can fill open cybersecurity positions is a growing problem, both in the U.S.A. and globally. There is a clear need to grow the number of people who can fill these positions. Building from our experience collaborating with a nearby community college, we have developed a distributable course of video lectures that are designed for community college students with the goal of increasing interest in the career field of cybersecurity. The course is composed of four modules: offensive security and digital forensics, the internet of things, cryptography, and blockchain technologies. Each module consists of three or four video lectures delivered by JHU faculty, staff, and graduate students, along with accompanying reading material and exercises. The course is being made freely available to anyone who would like to have access to it. It has been designed to be put to use in numerous ways including, as a course for credit, a seminar series, individual standalone lectures or modules, and for use as content to complement existing curriculum.

Keywords: Cybersecurity · Education · Curriculum

1 Introduction

It is almost common knowledge that there is a serious shortage of people who can fill positions in cybersecurity. According to Cyberseek, a public private partnership between Burning Glass Technologies, CompTIA, and The National Initiative for Cybersecurity Education (NICE), from October 2018 through September 2019, there were over 500,000 open cybersecurity positions and just under 1 million people currently employed in the field indicating a supply/demand ration well below the national average [1]. Prior data from Cyberseek indicated that there were approximately 313,000 postings for cybersecurity positions [2] indicating that demand is increasing.

These numbers are for the United States alone. In 2017, The Center for Cyber Safety and Education (a "501(c)(3) segregated fund of International Information Systems Security Certifications Consortium, Inc." a company that provides training and industry recognized certifications) released its "Global Information Security Workforce Study" where they predicted a global shortage of cybersecurity workers of 1.8 million by 2022 [3]. The report also states that in 2015, Frost and Sullivan, a contributor to the report,

K.-K. R. Choo et al. (Eds.): NCS 2020, AISC 1271, pp. 37–45, 2021.
https://doi.org/10.1007/978-3-030-58703-1_2

estimated that the global cybersecurity worker shortage would be 1.5 million workers by 2020 [3].

Clearly there is a growing global need for skilled cybersecurity workers. As part of this need, it seems self-evident that there is a need to find ways to attract people to the field of cybersecurity from wherever they can be found. In this paper we outline one of our efforts to increase interest in the field by developing a freely distributable curriculum targeted at community college students. The initial goal of the course is to provide an overview of the of some interesting areas in cybersecurity to encourage new students to enter the field and existing students to continue on to a four year degree and beyond. This initial goal remains the main focus of the course, however this paper outlines additional goals we have discovered and embraced.

We aim to contribute to cybersecurity learning resources by providing a set of learning modules that cover both fundamental theories, e.g., cryptography, and emerging technologies, e.g., blockchain and Internet of Things security. In Sect. 2 we will discuss the background and why we were well-positioned to develop this curriculum. In Sect. 3 we will discuss the process we used for developing the curriculum. In Sect. 4 we will outline the curriculum and how it can be used. In Sect. 5 we will outline how the curriculum can be accessed for use in any setting.

2 Background

In 2015, The Johns Hopkins University Information Security Institute (ISI) partnered with Hagerstown Community College (HCC) in Hagerstown, Maryland to deliver a series of seminars on topics in cybersecurity. ISI faculty, staff, and students delivered lectures to HCC students on topics including digital forensics, actively attacking commercial unmanned aerial systems (drones), digital signatures, and fully homomorphic encryption. Some lectures covered the speaker's own research, and others covered well researched topics.

ISI delivered the lectures to HCC in the fall of 2015, 2016, and 2017. The majority of the lectures were delivered remotely via teleconferencing tools. While ISI personnel developed the lectures and the topics of discussion, any assignments were developed by HCC personnel.

As the collaboration with HCC was coming to a close, we began thinking about ways we could build on our experience. We envisioned a course modeled on our success with HCC but having more impact and the ability to reach a much wider audience. Additionally, we envisioned a course that would expose community college students to a broad range of cybersecurity topics with the goal of sparking new interest or increasing existing interest, inspiring students to enter or go further into the field and possibly seek more advanced degrees or careers in cybersecurity.

Our experience with HCC, while successful, was time consuming. It would be time, labor, and cost prohibitive for us to try to reach a larger audience with the model we used with HCC. To put it simply, the solution doesn't scale. Thus we envisioned a distributable course package that we could make available to community colleges. The course package would include a video lecture, some background information on the topic of the lecture, and assignments or lab exercises.

We anticipated a wide range of ability and interest among community college students. To that end we decided that each lecture would include exercises of varying degrees of difficulty ranging from simple recitation of the material in the lecture to exercises that require significant additional effort including research and synthesis. We tried to develop exercises that the faculty could use to construct assignments tailored to their students' needs.

3 Curriculum Development Process

The first step in our process was to develop a list of lecture topics for the course. We put together a list of topics with the help of ISI professors and researchers and found that it could be broken down into three main areas: Offensive Security, Internet of Things, and Cryptography. We grouped our lectures into these three areas.

One of the worst things we could have done is to develop this course without collecting input from community college faculty. We asked for input and held a remote meeting with community college faculty from Hagerstown Community College, Anne Arundel Community College, Prince Georges Community College, and Wake Technical Community College to propose our course design and lecture topics. We received some very positive feedback from these faculty members. The key takeaways from the meeting were:

1. The faculty were very interested in blockchain. Our initial list of lectures included a single lecture on blockchain but the faculty we talked with wanted a much deeper dive into the blockchain technology.
2. Faculty may not be interested in delivering the entire set of lectures. They would like the course to be flexible enough to be able to deliver a single lecture.
3. The faculty were very interested in being able to use the lectures we developed as part of their existing curriculum. They would like to use the course as a way to supplement their current resources.

In response to the input, we made changes to our course. We adopted a more modular design philosophy. To the extent possible we isolated the content of the lectures so they could be delivered without the students having seen all the previous lectures. There are still some prerequisites for some of the lectures but we minimized these relationships as much as we could. We added a topic area for blockchain and removed some more advanced cryptography topics from the lecture series. Finally, in designing and developing the learning materials we took into consideration that community college faculty may want to augment existing resources with our lectures and exercises.

We began developing the course and filming lectures. In 2018 we met again, this time in person, with community college faculty to demonstrate a lecture and the usage of the materials. Overall the feedback was positive and helpful. In the lecture we demonstrated, the lecturer was standing in front of a large monitor with the slides being displayed behind him. We received comments that the slides were difficult to read. The participants recommended that the slides be the sole image on the screen and that the lecturer should not take up screen space. We took this advice and reedited the videos we had already made to fit this recommendation. We completed the development of the course with this additional input in mind.

4 Curriculum Contents

4.1 Overview

The course is broken into four main subject areas we call modules. They are Offensive Security and Forensics, Internet of Things, Cryptography, and Blockchain. Each module contains three to four lectures. The course is available as a collection of files. Each lecture is contained in a single directory (or folder) and includes the following items:

1. A lecture video (MP4 Video).
2. A note for the instructor that describes the contents of the folder and provides any necessary background information (Word Document).
3. A reading list of papers and articles that the students and the instructor may want to read before viewing the lecture video (Word Document).
4. Two or three assignments requiring increasing degrees of effort (Word Documents).
5. The answer keys for the assignments (Word Documents).
6. The slide deck from the video lecture (PowerPoint file).
7. Other files that may be needed for the assignments (various file types).

The assignments are exercises that the students can complete after viewing the lecture video. Assignment 1 is always a recollection exercise that any student should be able to complete after watching the video. Additional assignments will require one of two activities on the part of the students. Some assignments will require that the student do some independent learning or research to properly answer the questions. Other assignments will require that the student perform some task. The latter form of assignment is usually well-suited for use as an in-class lab exercise.

4.2 Module and Lecture Descriptions

Offensive Security and Forensics. This module covers some novel topics as well as some standard topics.

Consumer UAV Research. Dr. Lanier Watkins, from ISI and JHU/APL, covers some common vulnerabilities in consumer grade unmanned aerial systems (UAS) commonly referred to as "drones." He discusses and shows the results of his students' research on hacking into multiple consumer grade UAS using common exploits. He proposes a way to defend against the physical security threat that these UAS present by using the hard-to-patch vulnerabilities against rogue drones. Assignment 3 in this lecture is the analysis of an included packet capture file using the open source tool Wireshark. Wireshark is not included as it is open source, freely available, and updates frequently. This lecture can stand alone and has no prerequisite lectures. The length of the video lecture is 48 min.

Covert Channels. Side channels and covert channels involve the use of nonstandard data analysis to either reveal or transmit data in a manner that is difficult to detect due to its obscurity. In this lecture, Dr. Lanier Watkins provides an overview of side channels then discusses his research in using a side channel to both infer malware infections and to send covert messages. Assignment 3 in this lecture involves the analysis of timing data in a spread sheet to determine the threshold value for covert channel to properly function. This lecture can stand alone and has no prerequisite lectures. The length of the video lecture is 44 min.

Introduction to Computer Forensics. Dr. Timothy Leschke, from ISI, introduces the students to the topic of computer forensics. The students are introduced to foundational legal material such as search warrants and the fourth amendment of the U.S. Constitution. Students are also introduced to the role of the forensic investigator and the responsibilities that go along with that role. Dr. Leschke covers the types of digital artifacts that may be used in court and how they are collected, protected, and ultimately provided as evidence in a trial. Assignment 3 for this lecture is an exercise in computer forensics. Students will be provided with disk image files (included with the lecture) and instructions on installing Autopsy, an open source forensic tool kit. Autopsy is not included as it is open source, freely available, and updates frequently. This lecture can stand alone and has no prerequisite lectures. The length of the video lecture is 34 min.

Passwords and Authentication. Joseph Carrigan, from ISI, covers the history of authentication from the days before passwords to modern time. The lecture discusses the evolution of password-based authentication from storing passwords in plain text through hashing and salting. The lecture covers how passwords are cracked and includes demonstrations of two powerful and popular password cracking programs, Hashcat and John the Ripper. Assignment three for this lecture is an exercise in cracking passwords. Students are provided with a "shadow" file from a Linux system. The passwords are salted and hashed with MD5. Students can use the file "password.lst" to crack all of the passwords but they will need to use some more advanced features of the password cracking program to crack all of the passwords. Students without access to GPUs should use John the Ripper. Students with access to GPUs can use either John the Ripper or Hashcat. Both of these applications are open source and freely available. This lecture can stand alone and has no prerequisite lectures. The length of the video lecture is 48 min.

Internet of Things. The Internet of Things represents one of the largest attack surfaces on the internet. Many of these devices are not properly secured and can't be secured for use on the open internet. This module introduces students to the concept of embedded systems, IoT Devices, and SCADA/ICS and the security issues that come with these technologies.

Embedded Systems Overview. Joseph Carrigan introduces the student to what an embedded system is and provides a real world example and comparison to general purpose computers. The student is introduced to the underpinning technologies of embedded systems including Application Specific Integrated Circuits (ASIC), Field Programmable Gate Arrays (FPGA), Microcontrollers, and Single Board Computers; next, the student is introduced to what embedded systems do. Finally the student is introduced to some common embedded system platforms that are available and guided through two examples of embedded systems, a people detector based on an Arduino and a security camera based on a Raspberry Pi. Assignment 3 in this lecture is an exercise in designing (but not implementing) a burglar alarm system. The lecture includes the Arduino code for the people detector. If the student has access to an Arduino and sensors they can implement something similar. This lecture can stand alone and has no prerequisite lectures. The length of the video lecture is 45 min.

Consumer Grade IoT Devices. David Halla from JHU/APL introduces the student to the Internet of Things (IoT), the Industrial Internet of Things (IIoT), and the trends and challenges of these fast-growing fields. The student is then introduced to the root causes and the implications of the security challenges facing IoT and IIoT implementations. Next, the student is introduced to some ways to address IoT security. Finally, Joseph Carrigan provides a real world demonstration of penetrating an IoT network that has some very common security vulnerabilities, i.e. default passwords, and shows how attackers can exploit this vulnerability and move around a network. We recommend that students view the "Embedded Systems Overview" lecture before watching this lecture, but this lecture could stand on its own. The length of the video lecture is 36 min.

SCADA and ICS Intrusion Detection. Dr. Lanier Watkins introduces the student Industrial Control Systems (ICS) and Supervisory Control and Data Acquisition (SCADA). The student is introduced to the severity of security compromise in ICS using Stuxnet as an example. Dr. Watkins introduces the student to some of his research on detection of malicious software on SCADA and ICS systems. Assignment 3 in this lecture involves the examination of results from various machine learning tools. We recommend that students view the "Embedded Systems Overview" lecture before watching this lecture, but this lecture could stand on its own. The length of the video lecture is 45 min.

Cryptography. This module introduces the student to the history and foundational concepts of cryptography. This module also lays the foundation for the necessary understanding to proceed to the Blockchain module.

Introduction to Cryptography Part 1. Dr. Matthew Green from ISI introduces the student to the basic concepts of cryptography. The lecture starts with understanding the basic communication model and the types of adversaries in communication. The student is introduced to the history of cryptography including classical cryptography such as substitution cyphers and offset cyphers. The student learns about one-time pads and why they are still unbreakable. The topic then moves to mechanical cyphers such as the Enigma. The lecture then moves on to modern encryption where the student is exposed to block cyphers, specifically the Data Encryption Standard (DES) and the Advanced Encryption Standard (AES). The lecture concludes with a discussion of hashing algorithms. Assignment 2 is the decoding of a Caesar cypher. This lecture can stand alone and has no prerequisite lectures. The length of the video lecture is 51 min.

Introduction to Cryptography Part 2. Dr. Matthew Green continues where he left off in the last lecture with a review of hashing functions and block cyphers. The student learns about block cypher modes and correct and incorrect usage of these modes. Next the student is exposed to Message Authentication Codes (MACs) and Authenticated Encryption. Finally the lecture delivers an introduction to Asymmetric Encryption. Students are exposed to Diffie-Hellman Key Exchange, Public Key Infrastructure (PKI), the RSA algorithm, and Digital Signatures. In assignment 3 for this lecture, the student is challenged to implement AES to encrypt the contents of a text file using a library. "Introduction to Cryptography part 1" is prerequisite for this lecture. The length of the video lecture is 55 min.

Advanced Topics in Cryptography. Dr. Abhishek Jain from ISI provides a more in-depth look at some of the topics in the "Introduction to Cryptography part 2." The student is exposed to group theory (cyclic groups) and hard problems (the discrete logarithm problem), two ideas that underpin key exchange protocols and asymmetric encryption. The student is shown how these ideas are implemented in Diffie-Hellman to provide an unforgeable secure communication channel. Assignment 3 for this lecture is a trivial manual walkthrough of the Diffie-Hellman key exchange. It is best suited for three students acting in the roles of sender, receiver, and eavesdropper. "Introduction to Cryptography part 1" and "Introduction to Cryptography part 2" are prerequisites for this lecture. The length of the video lecture is 39 min.

Blockchain. This module introduces the student to blockchain and its usage in cryptocurrencies and beyond. "Introduction to Cryptography part 1" and "Introduction to Cryptography part 2" are prerequisites for this module. Also, we recommend "Advanced Topics in Cryptography" as a prerequisite though it is not absolutely required.

Blockchain and Cryptocurrencies Part 1. Dr. Abhishek Jain provides a history of digital payment systems and their problems. Dr. Jain discussed the ideas of centralized and decentralized electronic cash and introduces the student to the first digital currency, Digicash, and why it failed. Next the student learns about Visa and MasterCard's Secure Electronic Transactions (SET) and why it failed. Then the student is introduced to Bitcoin and the important features of consensus and decentralized identity management. The student is then introduced to the concepts of cryptocurrencies through two example cryptocurrencies. Assignment 3 for this lecture asks the student to speculate on the true identity of Satoshi Nakamoto and support their conclusion. The length of the video lecture is 44 min.

Blockchain and Cryptocurrencies Part 2. Dr. Abhishek Jain continues the exploration of cryptocurrencies, in particular Bitcoin with a discussion of Bitcoin's consensus protocol. This discussion covers the difficulty of consensus and examines why Nakamoto's solution is so innovative. Next the student is exposed to Bitcoin's Proof of Work method for selecting a random node to add the next block to the blockchain and how this random selection contributes to the security of Bitcoin. Then Dr. Jain discussed the limitations of Bitcoin including functionality, privacy, and sustainability. In addition to the prerequisites for this module, "Blockchain and Cryptocurrencies part 1" is a prerequisite for this lecture. The length of the video lecture is 50 min.

Blockchain Beyond Cryptocurrencies. Gabriel Kaptchuk, a Ph.D. student from JHU Department of Computer Science, provides a review of some of the key concepts of

Blockchain technology. The lecture continues to a discussion other applications for blockchain technology. These applications include censorship resistant microblogging (tweeting), secure state maintenance, and fair multiparty computation. Each of these applications is explored with examples. Assignment 3 for this lecture asks the student to describe the selfish mining attack on proof-of-work networks. This will require the student to research the attack. Resources are provided. In addition to the prerequisites for this module, "Blockchain and Cryptocurrencies part 1" and "Blockchain and Cryptocurrencies part 2" are prerequisites for this lecture. The length of the video lecture is 1 h 9 min.

4.3 Suggested Uses

There are many ways that we envision this course being put to use. The course can stand on its own as a full course for a typical semester. There are 13 lectures that can fit into a standard 15 week semester with one week for orientation at the beginning of the semester and one week for review at the end of the semester. Each week students can review the required reading either in class or on their own, then in class view the lecture video. Students can either do the assignments as homework or work together in class to complete them. Some of these assignments lend themselves to in class work or laboratory exercises. Faculty could also take the assignments provided with the lectures and make one homework assignment from the provided content.

The course could be used as a seminar series where students simply view the lecture videos. Students wishing to know more or to have a better understanding can be provided with the exercises and reading materials.

The lectures that stand on their own can be presented as special topic events. The lectures which must be grouped together, in particular the cryptography module and the blockchain modules, can be presented as a shorter course or seminar series as described above.

Faculty are also welcome to take whatever they would like to take and add it to their existing curriculum as a supplement or as material for students to view on their own. This includes the lecture videos, the assignments, and the reading materials.

While this course was designed with community college students in mind, we will provide it to anyone with any interest in the materials, including high school faculty and faculty at four year institutions. The materials can be used in any way seen fit including adding some or all of them to existing curriculum.

5 Distribution and Metrics

The course is available via a website https://cybercourse.isi.jhu.edu/. A username and password will be provided to anyone who requests it. The course is available at no cost. We will ask that anyone one who requests a copy of the course provide us with some information about how they intend to use the course. We will also ask that they provide us with information on how the course materials were actually used. Finally, we will ask that if the course is delivered to students, in whole or in part, and not used to supplement existing curriculum, that the faculty distribute and collect a student questionnaire on the material presented. Providing or agreeing to provide any of this information will not be a requirement of being granted access to the course materials.

Acknowledgement. This work was supported by NSA CNAP S-004-2017 grant number H98230-17-1-0327.

References

1. Cybersecurity Supply/Demand Heat Map. https://www.cyberseek.org/heatmap.html. Accessed 26 Dec 2019
2. Cybersecurity Supply/Demand Heat Map retrieved from the Internet Archive. https://web.archive.org/web/20190901211106/
3. (ISC)^2, Booz Allen Hamilton: 2017 Global Information Security Workforce Study, Benchmarking Workforce Capacity and Response to Cyber Risk. https://www.iamcybersafe.org/s/gisws. Accessed 26 Dec 2019

Digital Forensics Education Modules for Judicial Officials

Ragib Hasan[1(✉)], Yuliang Zheng[1], and Jeffery T. Walker[2]

[1] Department of Computer Science, University of Alabama at Birmingham, Birmingham, AL 35294, USA
{ragib,yzheng}@uab.edu

[2] Department of Criminal Justice, University of Alabama at Birmingham, Birmingham, AL 35294, USA
jeffw@uab.edu

Abstract. As our lives become more dependent on digital technology, cyber crime is increasing in our society. There is now an ever-increasing need to counter cyber crime through digital forensics investigations. With rapid developments in technology such as cloud computing, the Internet of Things, and mobile computing, it is vital to ensure proper training of law enforcement personnel and judges in the theory and practice of digital forensics. In this paper, we describe our methods and approach to create curricula, educational materials, and courses for training law enforcement and judicial personnel in digital forensics. We partnered with legal experts to design a series of modules/courses on digital forensics to educate the actual target demographics. Training materials have been designed to be not only scalable to nationwide law enforcement and judicial professionals, but also amenable to regular updates to respond to rapidly changing attacks and forensic techniques.

Keywords: Digital forensics · Education

1 Introduction

In recent years, advances in computing has changed many aspects of our lives. The rapid growth of cyber technology has significantly improved different domains. However, at the same time, cyber crime is rising, leading to malicious use of computing technology. Prosecutors are increasingly relying on technology and digital forensics to investigate criminal activities. A report of the FBI states that, during the fiscal year 2012 alone, the Computer Analysis Response Team (CART) of the FBI supported nearly 10,400 investigations and conducted more than 13,300 digital forensic examinations that involved more than 10,500 terabytes of data [28].

The very nature of cybercrime and digital forensics is also changing as new technology is adopted by the society. For example, with the emergence of cloud computing, consumers are increasingly moving to the cloud for their storage

K.-K. R. Choo et al. (Eds.): NCS 2020, AISC 1271, pp. 46–60, 2021.
https://doi.org/10.1007/978-3-030-58703-1_3

needs. In 2012, Gartner predicted that consumers will store more than one third of their digital content in the cloud by 2016 [17]. By 2019, the use of clouds to store customer data has skyrocketed – a 2019 report by Gartner predicts that "by 2022, 75% of all databases will be deployed or migrated to a cloud platform, with only 5% ever considered for repatriation to on-premises" [15]. Because of the large scale migration to the cloud-based storage and computation services, a large amount of forensic evidence is now derived from the cloud. Some incidents of storing contraband documents in cloud-based storage systems have already been reported [9,12]. Evidence residing on clouds has great impact on legal rules and regulations [8,14,21,25]. To prosecute and litigate a crime today, judicial officials therefore need detailed and advanced knowledge of computing, especially in the area of computer security and digital forensics. With rapid developments in technology such as cloud computing, the Internet of Things, and mobile computing, it is vital to ensure proper training of judges and prosecutors in the theory and practice of computer security and digital forensics.

Unfortunately, most of current cyber security and forensics education are geared towards technical experts or law enforcement investigators rather than judicial officials such as judges and lawyers. Most, if not all, of computer security and forensics educational material assume prior knowledge of computing and technology basics. People with a non-computer science background have difficulty in utilizing such educational materials to get a working knowledge of computer security and forensics. As a result, the judges and other officials have to blindly trust the experts associated with the trial, and assess the evidence with incomplete knowledge of the domain. The lack of domain knowledge in computer security and forensics technology can, and often does, lead to miscarriages of justice. Often, the digital evidence forms the core of a case and therefore the judges need to fully understand various aspects of the forensic evidence rather than completely relying on the expert witnesses. Therefore, there is a significant and urgent need for domain specific and appropriate computer security and forensics educational materials for judicial officials.

In this paper, we present an overview of our ongoing work to develop curricula, educational materials, and courses for training law enforcement and judicial personnel in digital forensics. We have created a set of educational modules and courses of various lengths that are geared toward judicial officials. To do so, we have partnered with judicial officials working in the area to develop and disseminate the courses and evaluate their effectiveness in educating the target demographics. We have also included mechanisms to frequently update the educational modules to match the rapid growth of technology. Our target audience includes judicial officials, including judges, prosecutors, attorneys, investigators, and other judicial personnel.

Contributions: The contributions of this paper are as follows:

1. We explored various aspects of legal cases and judicial processes to identify the specific domain knowledge that judicial officials require for learning forensics.
2. We developed domain specific and appropriate educational modules to teach computer security and forensics to judicial officials. The modules have been

designed to range from self-paced online courses to a single day short course, or a series of multiple courses to provide a comprehensive knowledge.
3. We also developed a continuous improvement process to evaluate and improve the effectiveness of various dissemination mechanisms to deliver the modules to judicial officials.

Organization: The rest of the paper is organized as follows: in Sect. 2, we provide background information on computer forensics and relevant laws. Sect. 3 provides motivation for creating this domain specific educational resource for judicial officials. We discuss the challenges of this work in Sect. 4. We provide details of our approach towards developing the educational materials in Sect. 5, and provide a sample module syllabus in Sect. 6. Finally, we conclude in Sect. 7.

2 Background

To provide the readers with an understanding of the scope of our work, we begin by exploring background information on computer forensics and the various laws governing the use of digital forensics in legal cases.

2.1 Computer Forensics

Computer forensics is the process of preserving, collecting, confirming, identifying, analyzing, recording, and presenting crime scene information. Wolfe defines computer forensics as "a methodical series of techniques and procedures for gathering evidence, from computing equipment and various storage devices and digital media, that can be presented in a court of law in a coherent and meaningful format" [30].

According to a definition from NIST [19], computer forensic is "an applied science to identify an incident, collection, examination, and analysis of evidence data". In computer forensics, maintaining the integrity of the information and strict chain of custody for the data is mandatory. Several other researchers define computer forensic as the procedure of examining computer systems to determine potential legal evidence [20,23]. In recent years, the term Digital Forensics have become more widely used, since the forensic evidence can come from many electronic devices such as smart phones, GPS modules, etc., which are not usually considered as computers.

From the definitions, we can say that computer forensics is comprised of four main processes:

Identification: Identification process is comprised of two main steps: identification of an incident, and identification of the evidence, which will be required for successful investigation of the incident.

Collection: In the collection process, investigators extract the digital evidence from different types of media e.g.., hard disk, cell phone, e-mail, and many more.

Additionally, they need to preserve the integrity of the evidence.

Organization: There are two main steps in organization process: examination, and analysis of the digital evidence. In the examination phase, investigators extract and inspect the data and their characteristics. In the analysis phase, investigators interpret and correlate the available data to come to a conclusion, which can prove or disprove civil, administrative, or criminal allegations.

Presentation: In this process, investigators make an organized report to state their findings about the case. This report should be appropriate enough to present to the jury.

2.2 Legal Basis of Computer Forensics

We discuss the legal basis of computer forensics by discussing its use in criminal and civil litigation.

Criminal Litigation: Digital forensics in criminal litigation is mainly governed by the 1986 Computer Fraud and Abuse Prevention Act. Subsequent court decisions such as Daubert vs. Merrell Dow Pharmaceuticals established the rules for admissibility of digital forensic evidence. According to the Daubert ruling, digital forensic evidence must be subject to the following standard [16]:

1. "Testing: Has the scientific procedure been independently tested?
2. Peer Review: Has the scientific procedure been published and subjected to peer review?
3. Error rate: Is there a known error rate, or potential to know the error rate, associated with the use of the scientific procedure?
4. Standards: Are there standards and protocols for the execution of the methodology of the scientific procedure?
5. Acceptance: Is the scientific procedure generally accepted by the relevant scientific community?" [16]

Civil litigation: In United States, before 2006, there was no separate US Federal law for computer forensics investigation in civil cases. As computer based crime was increasing rapidly, the Advisory Committee on Civil Rules took initiative to resolve this issue at 2000. In 2006, the Federal Rules of Civil Procedure (FRCP) provided the groundwork for the practice of electronic discovery in rule 26(a)(1)(A), which is known as e-discovery amendment [1,18]. According to FRCP rule 34.(a), all Electronically Stored Information (ESI), including writings, drawings, graphs, charts, photographs, sound recordings, images, and other data or data compilations are subject to discovery by the litigating parties [2]. Transient data, including metadata, may also be considered as discoverable ESI [26]. Some important factors in the FRCP amendment, that are contributing in current digital forensics investigations are:

1. FRCP defines the discoverable material and introduces the term Electronically Stored Information (ESI). Under this definition, data stored in hard disk, RAM, or Virtual Machine (VM) logs, all are discoverable material for the forensic investigation.
2. It introduces data archiving requirements.
3. It addresses the issue of format in production of ESI. If the responding party objects to the requested format, then it suggests a model for resolving the dispute about the form of production.
4. It provides a Safe Harbor Provision. Under the rule of safe harbor, if someone loses data due to routine faithful operation, then the court may not impose sanction on him or her for failing to provide ESI [18,30].
5. A Litigation hold is known as a preservation letter or stop destruction request [26], which is introduced by this amendment. FRCP Rule 37 prevents an organization from removing documents from any of its storage system, which implies that ordinary data retention and cleaning policies should not be applied to ESI under a litigation hold [2,8].
6. There are also new and emerging challenges in forensics investigations in cloud environments [8,24].

3 Motivation

As we enter the third decade of the twenty-first century, there is an urgent need for training specifically for judicial officers. There are over 10,000 state and local judges in the US, and just under 3,000 federal judges. There are 2,300 prosecutor's offices at the state and local level, employing between 2 and 100 prosecutors. There are also 93 federal prosecutor's offices in the US, employing between 20 and 350 assistant prosecutors. This places the number of prosecutors between 5,000 and 20,000 prosecutors. With the increase in the involvement of digital evidence in both criminal and civil cases, it is likely many of these judges will have to rule on evidence of a digital nature in their cases. An October 2016 presentation from Joyce Vance, the erstwhile US Attorney for Northern District of Alabama, by 2020, almost all court cases would entail a cyber component [29]. The cyber component comes not only from the devices or computers used in the crime, but also to find more evidence and connections between various suspects.

The understanding required for judicial officials is different from that of law enforcement officers and investigators. Law enforcement must fully understand the technical aspects of computer forensic investigation. Judicial officials need a full understanding of the law related to digital evidence. These include the differences between digital and physical evidence under the law, potential negative influences on juries, how to address motions and challenges from both the prosecution and defense, and others. This means judicial officials need a different type of education related to digital evidence from those who conduct the investigations.

There are many computer forensic training courses for investigators, law enforcement officers, and students. For example, Zhang et al. have discussed digital forensics education at the undergraduate level through the use of experimental learning techniques [31]. In 2007, Choo et al. identified the future need for educating judicial officials about digital forensics [10]. However, there are only a few training opportunities for judicial officials; thus the need for educational materials and training outlined in this paper. Of the courses that are offered for judicial officials, many of them are residential and for extended periods, from a few days to three weeks [3,4]. They are also sporadic or not offered on a national scope. As an example, Clancy et al. at the National Center for Justice and Rule of Law at the University of Mississippi have organized Symposiums on the search and seizure of electronic information [11]. However, the main focus of these series of symposiums was on the seizure of information and 4th amendments. There is a need for a workshop focusing on all aspects of digital forensics. Finally, the number of judges and prosecutors that need this kind of training rapidly overwhelms current training options. Our current work would augment these courses and create a broader spectrum of courses available to judicial officers that would greatly improve their understanding of this areas. The increased understanding would provide for better legal decisions.

3.1 Differences Between Computer Forensics and Other Legal Issues

Computer forensic issues in court are complex and continually changing. Further, computer forensic issues have become a part of the procedure of criminal prosecution, not simply a point of fact. As a result, judges and prosecutors have to have a thorough understanding of computer forensic issues related to law. These involve, but are not limited to, seizure of digital evidence (computers, cell phones, GIS data from a variety of devices, social media information, etc.); how digital media is searched, stored, and presented in court; Fourth Amendment issues of rights to privacy; and a host of other computer forensic issues that may come up in a court proceeding.

What was once documents that were tangible artifacts (pictures, letters, contacts, etc.) when the Fourth Amendment was created are now almost universally contained in a digital environment. This represents the greatest challenge to the Fourth Amendment in the history of the US. As pointed out in Riley v. California [7], many of the issues faced by judges and prosecutors did not even exist with the Fourth Amendment was written. Not only must judges and prosecutors be trained in computer forensic issues, they must also continually update their training. Cyber security and cybercrime evolve so quickly, that their knowledge may be outdated within a few months.

Furthermore, judges cannot simply rely on expert witnesses as they have in the past. In many areas, such as medical procedures, there is a specific issue that must be addressed in the court proceedings (such as the effect of a drug or a cause of death). That is not the case for digital evidence, however. There is a much broader issue of this kind of evidence in the court proceedings. To allow a forensic examiner or an expert witness to serve as the source of knowledge in the digital evidence would be tantamount to allowing police officers to determine if a search was within the Fourth Amendment. Judges and other judicial personnel need a much better understanding of computer forensic issues and digital evidence.

3.2 Difference Between Training Judges and the Police

There is a great deal of training in computer forensics for police officers; but much less so for judicial officials [13,22]. There is also a significant difference in the kind of training necessary. Training for police officers is one of two types. The first is technical aspects of digital evidence. First, officers need to know how to safely and legally seize digital evidence and how to process it. This is very technical, hands-on training. Second, police officers need to know what the law says and how to work within the law. This is application of legal issues, where officers merely need to know what the law is and how they must act to comply.

Judges, however, are making the law. They must interpret the legal standards that have been set in higher courts; but, in the rapidly changing environment of computer forensics, this is not an easy task. Historically, application of the Fourth Amendment was fairly straight forward. There was a seizure of a tangible object (a letter, drugs, etc.). It was then a matter of interpreting the legal precedent related to seizures. This is completely different for digital evidence. First, there is the issue of whether information stored electronically is even physical evidence that falls under the Fourth Amendment. This is much more difficult than it sounds because, as soon as the law establishes the parameters for digital evidence in one device, technology changes and renders previous decisions moot (such as the difference between files stored on a hard drive and files stored in the Cloud). Judges must react to these changes within the short time frame of a trial, where interpretation of previous law may not be clear. For this reason, judicial officials need a strong training program so they can understand both the technical aspects of computer forensics and the legal issues and background. Also, this training must be ongoing to address the continual changes both in law and in technology.

3.3 Lack of Proper Understanding of Digital Evidence May Lead to Miscarriages of Justice

An example of the potential for miscarriages of justice related to digital evidence can be found in the 2014 Supreme Court case of Riley v. California, 134S. Ct. 2473 (2014) [7]. In that case, Riley was stopped and suspected of weapons violations. After arrest, officers searched Riley prior to moving him to pretrial detention. During an inventory of Riley's possessions, an officer accessed his cell phone and went through the information. One item the officer found was repeated use of a term that was related to a street gang. Riley's phone was examined at least one other time by officers. Based on the texts and images found on the phone, Riley was charged in connection with a recent shooting. At trial, Riley argued that the cell phone was protected under the Fourth Amendment, and that it should not have been searched without a warrant specifically connected to the shooting. The trial court denied the motion and Riley was convicted and sentences to up to life in prison.

The California Court of Appeals affirmed the conviction, relying on a previous ruling (People v. Diaz [6]), which had held that searches of cell phones by police were admissible in court if they were retrieved directly from the person arrested. However, the U.S. Supreme Court overturned the conviction. In its ruling, the Court ruled that, in its current state (as differentiated from previous cell phones that only served as phones) a cell phone was "not just another technological convenience" and that it holds the "privacies of life" (pictures, addresses, documents) that deserve the protection of the Fourth Amendment. In this case, the Court rejected a precedent that dealt with officer safety and destruction of evidence because the cell phone in this case was in the possession of the police and not accessible by Riley. The Court left open what it would rule in the case of a cell phone that a subject might be able to get to destroy evidence. The Court also left to future decisions whether accessing information that might be relevant to officer security (such as a text that the suspects confederates were headed to the scene to attack officers). The Court also wrestled with the difference between what officers are allowed to do in relation to seizures of physical evidence compared to digital evidence. This is something that will continue to be addressed in future cases. Finally, the Court discussed but did not rule on issues of the potential of another person to remotely wipe a cell phone and whether that would result in a need for officers to access and even forensically copy a cell phone. In this case, there were a number of hypothetical situations addressed; however, the Court chose to focus only on the narrow issue of the search of a cell phone related to an inventory search of an incarcerated person. This leaves open many issue that courts will have to wrestle with on an almost daily basis – it shows the potential for miscarriages of justice if courts do not make the proper interpretation, shows how important it is for judicial officials to have quality training in legal issues related to computer forensics, and demonstrates that this training must be continually updates so judicial officials understand both the technological advances and changes in legal thinking.

4 Challenges

Providing computer forensics education for judicial officials face several challenges.

4.1 Scope and Depth of the Curriculum

Computer security and digital forensics have become a very large field of knowledge. Judicial officials need to understand the latest techniques used in cybercrime investigations. However, currently available educational material and courses in security and forensics often requires technical knowledge of computing, which is outside the expertise of most judicial professionals. Therefore, the curriculum needs to provide in-depth coverage of the latest topics of computer security and digital forensics while not requiring deep technical expertise. The challenge is to determine the scope and depth of the curriculum and to present security and forensics concepts at the level understood by judicial officials.

4.2 Designing a Domain-Specific Curriculum

The curriculum also needs to be highly domain-specific. Many concepts in computer security and forensics are addressed and understood in different names and terms in the judicial/legal domain. Also, many terms used in court proceedings have unique and specific meanings which may not be obvious to computer science and forensics educators. Therefore, it is important to explain various security and forensics concepts using the terminology used in the legal profession.

4.3 Time Constraints for Education

Judicial officials such as judges are extremely busy. Therefore, the courses and modules need to be short enough so the target demographics can afford to set aside time to explore the modules. For example, most if not all judges will not be interested in a semester long course, but may have time for a few days of training in digital forensics and computer security.

4.4 Finding the Best Dissemination Mechanism

The curriculum should be flexible enough to be taught in-class, through correspondence, or via online. Finding the best possible dissemination mechanism is one of the objectives of this work.

4.5 Keeping the Content Up-to-Date

The field of computing changes rapidly. Therefore, it is essential to keep the material up to date. However, updating the course content especially the videos is difficult and time-consuming. To complicate the matter, to add new content to the video resources, it must be consistent with existing videos (e.g.., taught by the same instructor, or in the same format).

4.6 Scalability

The biggest challenge is scalability. There are tens of thousands of judges and judicial officials in the US. Teaching them on-site in any one or a few particular locations is not feasible. However, the curriculum is vital for the entire community of judicial officials. Therefore, we must determine a way to scale the curriculum so that it can cover the entire community of judicial officials.

5 Approach Towards Developing the Educational Materials

To develop develop educational material for judicial officials and overcome challenges stated in the previous section, we took a multi-step approach involving a set of tasks, starting with requirements analysis for educational material, design

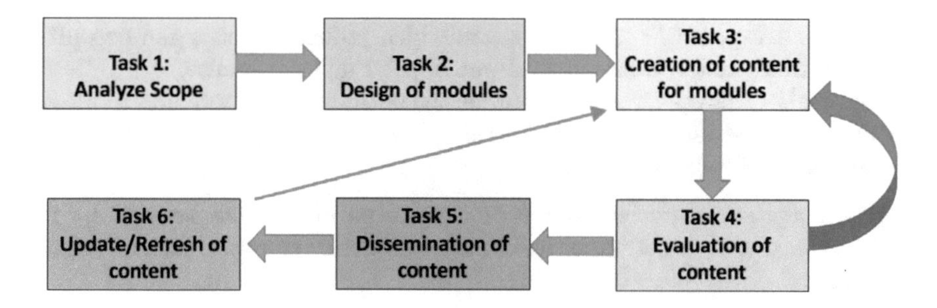

Fig. 1. An overview of the workflow for generating the educational modules

of customized modules, creation of the modules, the evaluation and dissemination of content, and periodic updates which focus on scalability and replicability. An overview of our workflow is shown in Fig. 1.

Next, we discuss each of the tasks in detail.

Task 1: Analysis and Customization of Educational Materials for Use by Judicial Officials. In this task, we analyzed and customize educational materials for use by judicial officials. The task involves two parts: collection of computer forensics educational material and discussion with judicial officials and justice sciences researchers to identify topics relevant and appropriate for judicial officials.

The goal of this task was to identify the subset of computer security and digital forensics educational material is relevant in the context of educating judicial officials.

Process: This phase was conducted in close collaboration between UAB's Computer Science and Criminal Justice departments. We have conducted meetings with judicial scholars to explore the research domain and create a set of topics that will be relevant to educate judicial officials. We also worked with lawyers to identify the topics.

Result: From Task 1, the result was a list of topics in digital forensics and computer security, ranked based on their significance. Also, we identified a list of foundational topics (i.e., basic terminology and concepts) that judicial officials without a technical background would need to know.

Task 2: Designing a Set of Modules with Different Paces and Learning Curves. Different judicial officials have different time constraints; therefore, we must create a flexible range of modules to suit all types of schedules. In this task, we created a set of educational modules with similar/overlapping curricula, covering digital forensics and computer science at various degrees of depth.

We have created multiple sets of modules at different levels. For example, this includes: 1) A one day crash course on digital forensics 2) A set of multiple day and multi-lecture modules for deeper exploration of computer forensics,

3) A set of self-paced learning resources which judicial officials can consult at their own time, and 4) a mobile and web app for quick reference.

The goal of this task was to determine the optimum set of modules to provide the best possible knowledge dissemination for a major portion of the target demographics of judicial officials.

Process: To determine how many different type of modules have to be prepared, we explored similar resources in other domains to get an idea of the best practices.

Results: The deliverable from this task were a set of module tracks, with various durations and depths, and a complete syllabus for each of the modules.

Task 3: Use of Video and Multimedia Technology to Create Modules. In this task, we created educational materials for the modules. This included development of video, web, and print materials to be distributed to the judicial officials.

The goal of this task was to determine the following (a) What is the best way to disseminate content to the target demographics? (b) What length of videos would be most preferred by the judicial officials?
Process: We used the following process to create the content:

- Video: For each module, we identified discrete topics and concepts covered by that module. We broke down the content into small chunks and short videos (5–10 min) on each topic. Breaking videos down to short chunks has several advantages: each chunk is easy to update without requiring edits to other chunks; short videos are also preferred by viewers when viewed online. Topics are relatively independent of one another to increase their chance of reuse in different modules. For the videos, we used both a classroom-based scenario (an instructor giving a lecture in front of a whiteboard) and a slide-based scenario (slides with narration), and a combination. We prepared the video lectures in accordance to the Quality Matters rubric for effective online courses [5].
- Text: We also created brief description of each topic along with examples. For each module, we prepared a short workbook.
- Web: Both the video and text material for the modules are hosted on a server to make them accessible to the target audience.

Results: The result from Task 3 were (a) a set of print materials, example problems, evaluation quizzes, (b) a set of videos covering various topics in the modules, and (c) a website hosting the modules and related links.

Task 4: Evaluation of the Quality of Modules. The goal of this task is to determine the quality of the educational material developed for this project and whether this will be effective for optimal dissemination of knowledge to the target demographics.

Process: In this ongoing task, we are working with UAB's Center for Educational Accountability (CEA) to evaluate the quality of the educational materials. The CEA evaluates a range of education, health, and training (e.g.., combat casualty training) programs.

The basic evaluation model to be used with this project is the Context, Input, Process and Product (Outcome) or CIPP evaluation model [27]. In this model, the context and input evaluations are designed for assessing the planning of the project and interpreting results, whereas the process and product evaluations are designed for assessing the implementation and outcome of the project.

Additionally, we plan to work with an educational consultant to evaluate the courses for Quality Matters (QM) certification [5]. We are also planning to make extensively use the IDEA survey of course participants to get feedback about the course.

Task 5: Dissemination of Modules to Target Demographics. In the ongoing Task 5, our goal is to explore different ways to disseminate the content to the target demographics. The various technique we have explored or plan to explore includes workshops offered at UAB Criminal Justice department, workshops at various state and national legal conferences for judges and prosecutors, partnerships with forensics standards bodies and organizations, through the website prepared in Task 3, and through a mobile app created especially for judicial officials.

Through these activities, we plan to determine the optimum, cost-effective, and most scalable method to disseminate the educational material to the target demographics.

Task 6: Periodic Update and Refresh of Modules. Here, we have identified the best practices for regularly updating and refreshing the module with new knowledge of technology used in digital forensics.

The goal of this task is to determine how we can keep the content up-to-date when the technology changes rapidly.

In accomplish this, we have developed a mechanism for efficient periodic updates to the modules. Technology changes rapidly, and we assume that we will have to update the content every year or every two years in order to provide an updated and current understanding of security and forensics. To do that, we have developed following workflow:

- During the update-review period, we will consult domain experts to determine whether the content is current and up-to-date.
- We will also crowdsource the analysis of the relevance of the modules by inviting the general judicial community to explore our modules and suggest changes.
- At the end of the review period, we will collect all the suggestions both from the experts and the crowd to determine which portions of the modules will require a change.

- We will then re-shoot the video segments, if possible with original narrators or teachers, and update the content. For new content, we will add them to the module repository. Also, the text content of the modules and the syllabus will be updated accordingly.
- For our mobile app, a push-notification will be sent to the mobile app to notify the user regarding the update to the content.

6 Sample Syllabus from the Educational Materials

Here, we provide a sample syllabus from one of our modules to demonstrate the structure of our educational materials.

Module 101: A One-Day Module on Computer Security and Forensics for Judicial Officials

Total hours: 5 h
Syllabus:

1. **Hour 1**: Basic security building blocks and terminology, common attacks, common defensive measures
2. **Hour 2-3**: Computers forensics, tools, steps, terminology, reporting rules, laws regarding computer forensics and digital evidence
3. **Hour 4**: Security and forensics in emerging technologies such as clouds, Internet of Things, mobile devices (smartphones, notebooks, tablets)
4. **Hour 5**: Discussion/Q&A/best practices/review

This module has been created for both an in-class and an online audience. For the latter, each hour has been broken into many video units, with each unit discussing a separate topic/concept. The last hour of discussion and Q&A for online students is done in the flipped mode, where an online live session is arranged monthly using Google Hangout.

7 Conclusion

In today's world, almost every aspect of our lives increasingly involves the use of computer technology. It is therefore imperative to educate the judicial officials about digital forensics process and best practices. In this paper, we have presented our approach towards creating a set of scalable and sustainable educational materials for teaching digital forensics to judicial officials.

We posit that the presence of such a set of educational materials would be highly beneficial to ensure proper education of judicial officials, which will lead to better and informed prosecution of legal cases. Educating the law enforcement and judicial personnel would allow them to understand and handle the increasingly omnipresent digital forensic evidence in legal cases and investigations. This would lead to significant improvements in investigating and prosecuting cybercrime, leading to a safer society for all.

Acknowledgements. This research was supported by the National Science Foundation through awards DGE-1723768, ACI-1642078, and CNS-1351038.

References

1. Federal Rules of Civil Procedure Rule 26. https://www.law.cornell.edu/rules/frcp/rule_26
2. Federal Rules of Civil Procedure Rule 37. https://www.law.cornell.edu/rules/frcp/rule_37
3. National Computer Forensics Institute (NCFI), Courses. https://www.ncfi.usss.gov/ncfi/pages/courses.jsf
4. National District Attorneys Association, Digital Technology Training. http://www.ndaa.org/digital_technology_training.html
5. Quality Matters Rubric. https://www.qualitymatters.org/why-quality-matters/about-qm
6. People v. Diaz, 51 Cal. 4th 84, 244 P. 3d 501 (2011)
7. Riley v. California, 134 S. Ct. 2473 – United States Supreme Court (2014)
8. Araiza, A.G.: Electronic discovery in the cloud. Duke L. Tech. Rev. 1 (2011)
9. BBC, Lostprophets' Ian Watkins: 'tech savvy' web haul, December 2013. http://www.bbc.com/news/uk-wales-25435751
10. Choo, K., Smith, R., McCusker, R.: "future directions in technology-enabled crime: 2007–09." "Research and public policy series no. 78. Canberra: Australian Institute of Criminology" (2007)
11. Clancy, T.K.: National center for justice and the rule of law (2004–2014). https://olemiss.edu/depts/ncjrl/index.html
12. Dist. Court, SD Texas. Quantlab technologies ltd. v. Godlevsky. Civil Action No. 4: 09-cv-4039 (2014)
13. Dotzauer, E.: COE - Cybercrime Training for Judges and Prosecutors: a Concept
14. Dykstra, J., Riehl, D.: Forensic collection of electronic evidence from infrastructure-as-a-service cloud computing. Rich. JL Tech. **19**, 1 (2012)
15. Feinberg, D., Adrian, M., Ronthal, A.: The future of the DBMS market is cloud (2019). https://www.gartner.com/document/3941821
16. Garrie, D.B.: Digital forensic evidence in the courtroom: understanding content and quality. Nw. J. Tech. Intell. Prop. **12**, i (2014)
17. Gartner Inc.: Gartner says that consumers will store more than a third of their digital content in the cloud by 2016 (2012). https://www.gartner.com/newsroom/id/2060215
18. K & L Gates: E-discovery amendments to the federal rules of civil procedure go into effect today, December 2006. http://www.ediscoverylaw.com/2006/12/articles/news-updates/ediscovery-amendments-to-the-federal-rules-of-civil-procedure-go-into
19. Kent, K., Chevalier, S., Grance, T., Dang, H.: Guide to integrating forensic techniques into incident response. NIST Spec. Publ. **10**(14), 800–86 (2006)
20. Lunn, D.: Computer forensics-an overview. Sans Institute **2002**, (2000)

21. Nicholson, J.A.: Plus ultra: third-party preservation in a cloud computing paradigm. Hastings Bus. LJ **8**, 191 (2012)
22. Proia, A.A., Simshaw, D.: Cybersecurity and the legal profession. Cybersecur. Our Dig. Lives **2**, 119 (2015)
23. Robbins, J.: An explanation of computer forensics. Natl. Foren. Center **774**, 10–143 (2008)
24. Ruan, K., Carthy, J., Kechadi, T., Crosbie, M.: Cloud forensics: an overview. In: 7th IFIP International Conference on Digital Forensics (2011)
25. Smith, J.: Electronic discovery: the challenges of reaching into the cloud. Santa Clara L. Rev. **52**, 1561 (2012)
26. Stacy, S.: Litigation holds: ten tips in ten minutes (2014). https://www.ned.uscourts.gov/internetDocs/cle/2010-07/LitigationHoldTopTen.pdf
27. Stufflebeam, D.L.: The CIPP model for evaluation. In: International Handbook of Educational Evaluation, pp. 31–62. Springer (2003)
28. The Federal Bureau of Investigation: Piecing together digital evidence (2013). https://www.fbi.gov/news/stories/2013/january/piecing-together-digital-evidence
29. Vance, J.: Partnering with the U.S. Attorney to Fight Cyber Crime. Cyber 2020, University of Alabama at Birmingham, October 2016
30. Wiles, J., Cardwell, K., Reyes, A.: The Best DAMN Cybercrime and Digital Forensics Book period. Syngress Media Inc. (2007)
31. Zhang, X., Choo, K.K.R.: Digital Forensic Education: An Experiential Learning Approach, vol. 61. Springer (2019)

Gaming DevSecOps - A Serious Game Pilot Study

James S. Okolica, Alan C. Lin(✉), and Gilbert L. Peterson

Air Force Institute of Technology,
2950, Hobson Way, WPAFB, Dayton, OH 454333, USA
{james.okolica.ctr,gilbert.peterson}@afit.edu, alan.lin@us.af.mil

Abstract. Serious games provide a method for teaching students in an engaging, non-threatening way. To be useful educational tools, these games need to be designed with the lesson objectives in mind. This paper uses the Game Design Matrix to design a game from learning objectives. The lesson objectives focus on software development, security and operations (DevSecOps). DevSecOps requires judging the cost and risk trade offs between secure and unsecured software aspects including code, data and usability; devising a strategy to produce a system under threat; comprehending that partial security may equate to no security; and identifying examples of insecure code and methods of mitigating it. These lesson objectives combined with the teaching environment drive design decisions that produce a game that fits well into a graduate level course in secure software design and development. We evaluate the game in a graduate course on secure software design and development. Learner surveys confirm that DevSecOps conveys the learning objectives in an engaging way.

Keywords: Serious games · DevSecOps · Game design methodology · Pilot study

1 Introduction

This paper introduces a serious game that teaches DevSecOps. Often seen as a leisurely pastime, games have emerged as an effective teaching tool [6], providing students the opportunity to learn without lectures. The game DevSecOps [21] is a part of a ten week graduate level course on secure software design and development. The course learning objectives include identifying security vulnerabilities in software and design specifications and writing secure software and design specifications. The game's learning objectives is to place secure software design and development within the context of the DevSecOps framework. The game is designed to be used twice. On the first day of the course, it is used as an introduction of the terms and concepts discussed in the class and where they fit

The views expressed in this paper are those of the authors and do not reflect the official policy or position of the United States Air Force or the Department of Defense.

K.-K. R. Choo et al. (Eds.): NCS 2020, AISC 1271, pp. 61–77, 2021.
https://doi.org/10.1007/978-3-030-58703-1_4

within DevSecOps. The second time is during instructor led review for the end of course exam. In this case, while it is again used to frame where the course material fits within the DevSecOps framework, it is also used as a review tool for the material on the exam. More formally, the game's learning objectives are:

1. Judge the cost and risk trade offs between secure and unsecured software aspects including code, data and usability.
2. Devise a strategy to produce a system under threats.
3. Comprehend that partial security may equate to no security.
4. Identify examples of insecure code and methods of mitigating them.

Using the Game Design Matrix (GDM) [22], the authors choose game dynamics, mechanics and components that allow the first three lesson objectives to be taught implicitly while the fourth lesson objective is taught explicitly. The game is designed to be flexible enough to allow players to explore different strategies with the hope that they will settle on one that maximizes the chance of successfully producing a system under threat. The game is also designed to be simple enough that players can learn it with minimal assistance. The hope is that players will take the game home and play it with friends and family, learning without focused intent.

A pilot study was conducted at the conclusion of a graduate level course on secure software design and development. Students were introduced to the game, provided 5 min of instruction and then left to play the game. After a single play, the students completed a survey consisting of a mix of long and short answers. Based on the survey results, 88% of the students said they either learned about the need to secure software from unexpected events (game learning objective 3) or that they learned about specific methods for mitigating software vulnerabilities (game learning objective 4) or both. Students employed a variety of successful strategies for the game (game learning object 2). Lastly, while high variance in inter-rater reliability prevents a quantitative analysis of whether students judged the cost and risk trade offs between secured between unsecured software aspects (game learning objective 1), student written responses like "the numerous ways to assist with gaining progress points show that there is no clear cut process to create secure software" suggest that some of the students learned that lesson as well. Students were neutral on how enjoyable and easy the game is to learn, suggesting that more game development is needed.

2 Related Work

2.1 DevSecOps

Software as a Service (SaaS) changes the traditional model of software development companies from creating a product that is sold and delivered to a customer to a model where the product is hosted at the development company and customers connect to and use the software without physically receiving it. This new paradigm facilitates software companies continuously integrating (CI) new development and continuously delivering (CD) it to customers since there is now a single copy of the software [20]. CI and CD introduce new security concerns. Security must be addressed from the beginning and continuously through the software's lifecycle as new features never end. DevOps is described as the "conceptual and operational merging of development and operations' needs, teams and technologies." [20]. Components of this merging include code, data, and usability along with the infrastructure that facilitates its effective creation.

DevSecOps adds the security to DevOps. As McGraw [18] observes, secure software is an emergent property of writing secure code. Howard [10] cites 24 deadly sins of software security including coding issues like buffer overflows and command injection as well as usability issues like weak passwords and usability. Addressing these issues throughout the continuous development process is the first step to DevSecOps.

2.2 Serious Games

The term serious game was first coined by Abt [1] in 1970 when he observed that "we are concerned with serious games in the sense that these games have an explicit and carefully thought out educational purpose and are not intended to be played primarily for amusement." While not designed primarily for fun, in a survey of game developers, educators, and researchers, over 80% of respondents felt that fun with serious games was either important or very important [19]. Michael and Chen argue that the main reason for instilling fun in games is to get people to play and learn on their own. Breuer and Bente [3] also observe that serious games may involve stealth learning where players become aware that they learned something useful only after they've played the game. Where these serious games excel is at teaching relationships, not facts [6].

As a society, we have been socialized that losing at games is expected. Therefore, when games are brought to the classroom, they provide teaching opportunities without stigma [6]. As Gee [8] observes, learning occurs when lecturing is kept to a minimum and the student is given the maximum opportunity to experiment and make discoveries. Farber [6] observes, "the content both at the surface level and at the gameplay level is why people play. What comes with it is what you actually teach. If you want to teach a particular thing, find something where knowing that particular thing will help you, and not knowing it will not hinder you, but it will slow you down a bit." Game designers search for combinations of game mechanics that accomplish this very thing.

Historically, game designers learn their trade over many years either by trial and error or by formal or information apprenticeship with an experienced game designer. For instructors to use serious games as educational tools, game design must be simple and straightforward. Several systematic game design frameworks have been developed [2,5,11,25]. Unfortunately, assessment of these game design frameworks is spotty, with several presented with at most one case study.

As shown in Fig. 1, the Game Design Framework (GDM) [22] begins with the designer defining the education objectives including learning objectives, constraints associated with the classroom environment, and the game domain

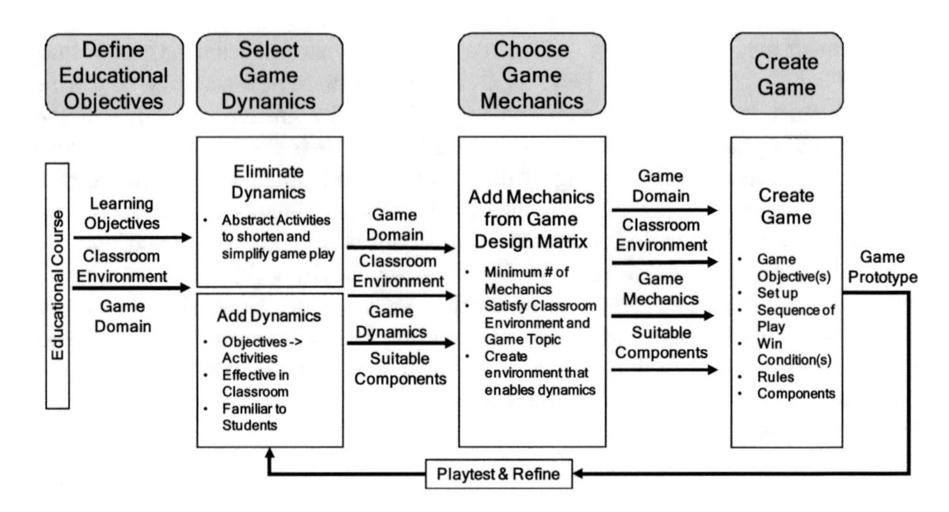

Fig. 1. Game design matrix [22].

(i.e., the class material simulated in the game). These inputs are used to select potential game dynamics, the system behavior of the game [13], that produce the "feel" of the game and create situations where applying the learning objectives helps to win. This is the most nebulous step in the GDM process. The selected game dynamics are then cross referenced in GDM's matrix with potential game mechanics, i.e., player actions, which are instantiated in the rules of the game. Once the designer has selected game mechanics, he selects game components and write down rules. GDM then recommends iteratively playtesting and refining dynamics, mechanics, and rules until the game meets the educator's objectives.

While there have been many serious games that focus on cyber security(e.g., [4,9,12,14,17]), very few are focused on secure software design and development and the authors failed to find any that focus on DevSecOps. Elevation of Privilege [23] is a card game used at Microsoft to help developers get better at finding software vulnerabilities in code. EoP is an unusual case. It is a card game in an environment where most serious games are video games and it is focused on secure software design where most cyber serious games focus on cyber security. EoP's small footprint and portability make it ideal for handing out at tradeshows and for use as a marketing tool as well as an educational one. EoP is also prefaced on a mixed group of players. Individuals more experienced with threat modeling and secure software design and development guide other players at discovering vulnerabilities and methods for mitigating them.

3 Transforming Learning Objectives into Game Mechanics

This section discusses the process by which the DevSecOps game was designed. First, the learning objectives and class constraints are discussed. Next, the game dynamics that support the learning objectives are covered followed by the game mechanics that support those dynamics. Finally, the section concludes with a discussion of our first presentation of the game to a group of instructors to identify issues and validate the capture of the learning objectives.

3.1 Learning Objectives and Class Constraints

The Game Design Matrix (GDM) [22] is used to design the DevSecOps game. The intention is for the game to be used twice in a ten week graduate level course on secure software design. The course learning objectives are:

- Identify security vulnerabilities in software and design specifications
- Write secure software and design specifications
- Develop the ability to quickly read and critically analyze peer reviewed papers
- Identify and succinctly describe key characteristics of a specific lesson

The course begins with an introduction to DevSecOps as a means for motivating why secure software design and development is so important. The first time that the DevSecOps game is used is during the first day of class. Its goal is to provide an opportunity for students to see a "real world" example of why security needs to be baked into the software development process rather than be bolted onto the end, introduce secure software design principles to the students (course objectives 1 and 2), and to show the students how a lecture on DevSecOps can be boiled down to a handful of key characteristics (course objective 4). The second time the game is used is near the end of the course as students are preparing for the course exam. At that time, the goal of playing the game is to remind students of the material learned in the course and to again remind them of the real world reason for designing secure software.

The lesson objectives for the game went through several iterations. Initially, it was about boiling down a ten week course into a 15 to 20 min game. Then, common sense re-vectored the game scope. Identifying and succinctly describing key characteristics is learned tangentially as part of the act of playing the game. Identifying and writing secure software and design specifications is simplified down to exposing the students to names and brief descriptions of topics covered in the course. What is left is a game that focuses on the lesson of the first day, DevSecOps. As a result, the learning objectives for the game are:

1. Judge the cost and risk trade offs between secure and unsecured software aspects including code, data and usability
2. Devise a strategy to produce a system under threats
3. Comprehend that partial security may equate to no security
4. Identify examples of insecure code and methods of mitigating it

As can be seen, the first three learning objectives focus on DevSecOps and the importance of security in the agile development cycle. Only the last game learning objective relates to the course objectives. As an introductory graduate level course in secure software design, the course includes exposing the students to new topics (e.g., identifying and understanding). However, since it is taught at the graduate level, the primary focus in on higher levels of Bloom's revised taxonomy [15] (e.g., judging and devising).

There are several constraints placed on the game. It needs to be played in a short period of time. Since several learning objectives are "baked in", students need to play the game multiple times. Given class time constraints, the game needs to play quickly. Play time between 15 and 20 min seems ideal. Additionally, class size is approximately 25 students. Creating a game that supports 3 to 4 players means that game components need to be inexpensive enough that producing 5 or 6 copies of the game is not cost-prohibitive. Currently, the class is taught in a lab where table space for game play is limited. Also, the rules need to be simple enough that the game can be taught quickly to people who may have not experienced board or card games more complicated than Monopoly, Scrabble, and Checkers.

3.2 Game Dynamics and Mechanics

Once the educator/designer has determined the learning objectives and constraints, the next step in GDM is to use them to decide on game dynamics. Game dynamics are the system behaviors of the game [13]. Pendleton [22] catalogs different game dynamics for different levels of mastery. From this list, several game dynamics from the evaluating/analyzing level of mastery are selected including multiple strategies and resource scarcity. Given the time and ease of learning constraints, the number of dynamics is kept to a minimum.

After choosing game dynamics, the next step in GDM is to select game mechanics that facilitate the game dynamics. Game mechanics are methods invoked by agents to interact with the game state [24]. Our goal is to create an environment where players have space to try several different viable strategies and judge which are more effective. Strategies include various levels of focus on security and infrastructure versus racing to complete a project before the attacks come (e.g., high risk/low security, low risk/high security, balanced, change from low to high risk, change from high to low risk, bolster infrastructure for fast and secure development). These strategies require measures of progress toward creating working software, measures of security levels, and measures of infrastructure. For security levels to be meaningful, there needs to be some negative consequence for not having sufficient security and for infrastructure to be meaningful, there needs to be some positive consequence for having it. Both of these suggest an environment where resources are scarce and must be "spent" to achieve these things. For the two dynamics of multiple strategies and resource scarcity, one common mechanic is transforming. Two others are allocating (multiple strategies) and buying/selling (resource scarcity). This combination of mechanics

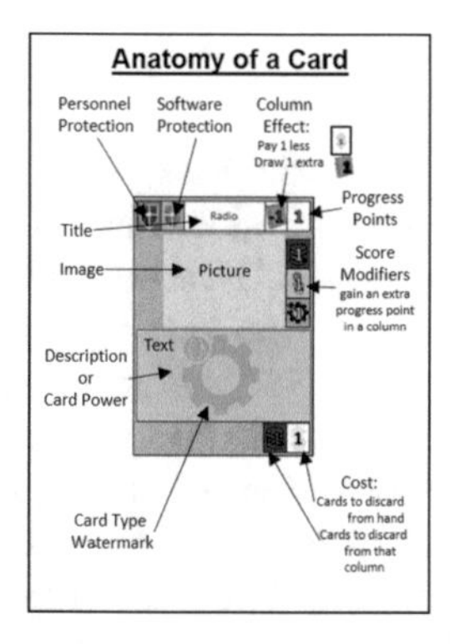

Fig. 2. Card anatomy.

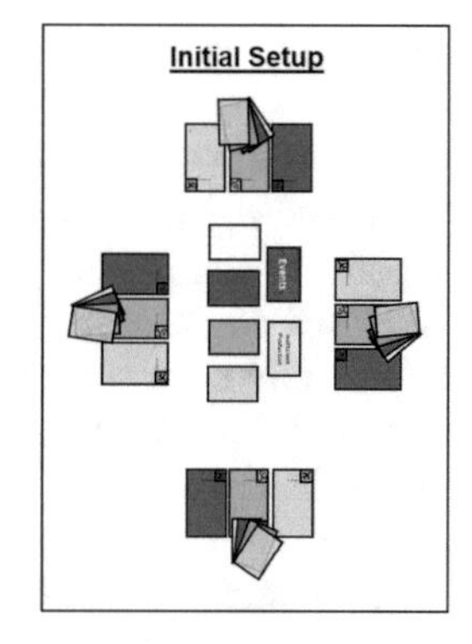

Fig. 3. Initial setup.

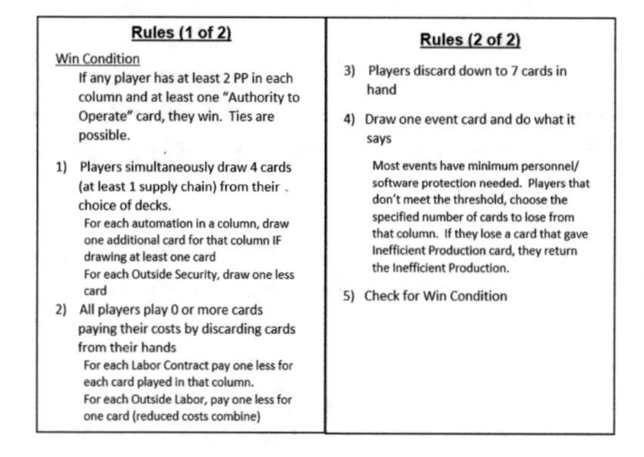

Rules (1 of 2)

Win Condition

If any player has at least 2 PP in each column and at least one "Authority to Operate" card, they win. Ties are possible.

1) Players simultaneously draw 4 cards (at least 1 supply chain) from their choice of decks.
   For each automation in a column, draw one additional card for that column IF drawing at least one card
   For each Outside Security, draw one less card
2) All players play 0 or more cards paying their costs by discarding cards from their hands
   For each Labor Contract pay one less for each card played in that column.
   For each Outside Labor, pay one less for one card (reduced costs combine)

Rules (2 of 2)

3) Players discard down to 7 cards in hand
4) Draw one event card and do what it says
   Most events have minimum personnel/ software protection needed. Players that don't meet the threshold, choose the specified number of cards to lose from that column. If they lose a card that gave Inefficient Production card, they return the Inefficient Production.
5) Check for Win Condition

Fig. 4. Rules on a single doublesided card.

suggests some method of acquiring resources by allocating actions or workers and then transforming those resources into other resources, possibly via buying or selling.

While GDM provides a list of dynamics and a matrix of dynamics and mechanics, it is silent on components. Based on the authors' familiarity with other games, including Race for the Galaxy [16], the decision was made to use multi-use cards (Fig. 2). These cards function as progress points toward creating a functioning piece of software; they function as security measures to securing the software from attacks; they function as infrastructure to make the development process more efficient; and they function as resources that can be used as payment for playing cards. There are three different types of multi-use cards corresponding to development, data, and usability. Once the decision to use multi-use cards is made, the rest of the initial design comes together quickly. The only other components in the initial design are event cards that are flipped each round to simulate attacks.

4 Rules Summary

Students are introduced to the game with some thematic background, "The DoD just released requirements for a 'next generation' AI pilot. First company to demonstrate a completed prototype gets the contract. Everyone else gets nothing. You are one of one of four defense contractors who think you've 'got what it takes'. Given the public interest, your progress will be reported each month in three separate areas: development, usability, and data components. Your progress in these areas will be labeled 'Progress Points'. Two progress points in each area, along with the ability to get it assembled and tested and you've got the contract... Good Luck!" As shown in Fig. 3, each of the different areas (development, data, and usability) has its own deck of cards (purple, orange, green). There is also a separate deck of Infrastructure cards (gray).

As shown in Fig. 4, the rules fit on a single double-sided playing card. The game is played in a series of rounds, each of which has four phases. Play continues until someone meets the win condition in phase four of a round. Phases occur sequentially, but players perform a single phase simultaneously (i.e., all players perform Resource Production then all players perform Project Advancement, etc.). The phases in a round are:

- Resource Production: players simultaneously select 4 cards from any combination of decks (e.g., 2 usability cards, 1 infrastructure card, and 1 data card). The only caveat is that each player must take at least one infrastructure card.
- Project Advancement: players play any number of cards from their hand to their tableau. Each card has a cost associated with it. To play a card, a player must discard cards from his hand equal to the cost of the card.
- Global Events: A card is turned up from the event deck that affect all players and generally involves losing cards in their tableau.
- Check for Project Completion: If a player has 2 Progress Points in each area and an "Authority to Operate" card in their tableau, that player wins (ties are possible).

4.1 Design Decisions

The three component decks represent management focus. For instance, drawing cards from the usability deck represents management's interest in advancing usability. However, it is the project advancement phase when players actually commit resources to these areas (by paying to play a card by discarding other cards). So, management may initially be focused on usability (2 cards) with some interest in data (1 card) and a required interest in infrastructure (1 card). However, when management meets to review ideas, they see advancing data will better allow them to achieve their goals. They then sacrifice the time they've spent on usability (discarding the 2 cards) to make progress in data (play the 1 card drawn).

In addition to their cost, component cards (technologies) provide benefits. Some cards provide measurable progress to completing the project (Progress

Points). Other cards do not specifically advance the project, but they make the particular component more secure. Each round, an event occurs that generally checks how much players (companies) have secured their areas and impacts those that have not secured them sufficiently. The game considers two types of attacks. Technological attacks represent hostile actors hacking into the component area's technology and either degrading, destroying or stealing it. Personnel attacks represent hostile actors enlisting individuals with inside access to either knowingly or unwittingly accomplish the same goals. To combat these attacks, there are cards in the component decks that provide security against these attacks. Players must decide whether to attempt to quickly advance their project without investing in security or to not spend any resources on project advancement until the component areas are secured. Most often, players will decide on a combination of the two, providing some security to some areas while taking chances in others.

In addition to secure and unsecured options, players learn about 2nd and 3rd order effects. Some component cards produce an effect in another area. For instance, choosing a "cutting edge API" for database access for development may slow (or reverse) progress in the data area. On the other hand, implementing error checking with information messages serves both usability and development, producing progress in both areas. Both of these effects are modeled through component cards that either add or remove progress from other component areas in the same company.

The game is designed from the outset with the lesson objectives and constraints in mind. The game components and mechanics are chosen to facilitate a learning methodology where students implicitly acquire the lesson objectives. For instance, there is no rule that says "protect your areas from outside attacks". However, when events occur that penalize players if their areas are not protected, players learn to protect those areas. The remainder of this section covers the different design decisions that were reached based on the lesson objectives. They fall into the following categories:

- **Brevity** Short play time (15–20 after the first play through), Teachable in 5 min, Compact form factor
- **Abstraction** Abstract the domain; sacrifice realism for speed and simplicity of play; Bake the lessons into the game mechanics
- **Cards** Cards as the only game component; Multiple card types provide dynamic focus; Cards as currency
- **Racing** Winner is the first to victory condition; Events that slow people down to balance speed vs. preparedness

4.2 Design Decisions 1: Brevity

The game is designed as a Day 1 introduction to the subject matter. The instructor begins with a short overview of the course. They introduce the game which the students play for approximately 30 to 45 min and then use the lessons the students learned by playing the game to motivate the remainder of the lesson.

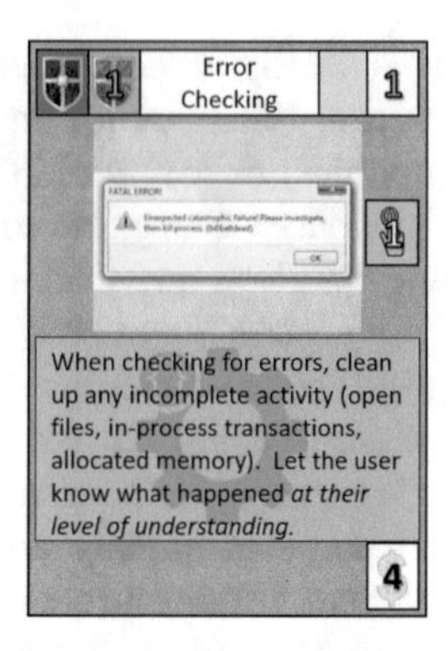

Fig. 5. Example of a Component Card.

A secondary objective is to provide copies of the game to the students so that they may play them in the evenings and after class. All of these conditions suggest creating a game that is easy to learn and quick to play. With a 15–20 min play time, the game can be played multiple times during Day 1 enabling the students to discover some basic concepts on their own. Furthermore, a short play time and simple components increase the chances of students bringing the game home and playing it in their free time, potentially exploring the learning space during non-class time.

4.3 Design Decision 2: Abstraction

Since the learning methodology is to enable the students to learn the lessons implicitly and explicitly, it is necessary to create a game that has one or more lesson objectives buried in the mechanics *and* has information embedded in the components (e.g., error checking as shown in Fig. 5). In the case of DevSecOps, this means providing secure software concepts and examples in the card descriptions and pictures. The game can initially be played focusing on the cost and risk tradeoffs in DevSecOps, without reading the descriptions or looking at the pictures, enabling lesson objectives one through three to be learned. Then, once the players have internalized the mechanics, they can look at the descriptions and pictures achieving lesson objective four while continuing to reinforce the first three objectives. The information in the cards that is communicated about secure software design and development is simplified into three component areas and an abstracted infrastructure area. This abstraction decreases the probability of students arguing about inaccuracies in the cards while still focusing them on the domain.

4.4 Design Decision 3: Cards

Given the cost nature of the domain, some sort of economy is necessary. Since the lesson objectives focus on trade offs, multi-use cards are used. Drawing cards represent management focus. Where is management currently focusing? Playing cards represent turning that focus into project progress. Discarding other cards to "pay" for the project progress represents the opportunity costs of not pursuing ideas that emerged during management focus. Differing costs for different cards represent the different amount of effort required for progress. Components with no security are inexpensive. Components with security and desirable 2nd order effects in other areas are expensive. Security procedures that do not provide progress fall somewhere in between. Finally, the choice of cards as the primary game component represents the uncertainty inherent in the development process. Sometimes, management wants to move in a certain direction but their employees' skill sets move them in a different direction. Balancing the card distribution within a deck as well as the number of decks demonstrates the uncertainty to the desired level.

4.5 Design Decision 4: Racing

Foundational to the lesson objectives is the trade off between security and cost. Developing a system in a secure environment is more expensive. Training personnel so they do not inadvertently provide assistance to adversaries takes time. Compensating them (financially or through other means) so they do not choose to assist adversaries is expensive. Companies must decide what the balance needs to be between security and cost. By providing a limited income each round (4 cards), players must consider trade offs between developing infrastructure first and simply trying to finish the job. Infrastructure includes security as well as infrastructure improvement cards. These improvement cards, found in the infrastructure deck, include benefits like discounts for playing cards and the ability to draw extra cards each round. Rather than having the game take a fixed number of rounds, it finishes as soon as a player has 2 progress points in each of the three component areas as well as a "Authority to Operate" card (found in the Infrastructure deck). This race mechanic forces the players to consider the trade off between infrastructure and speed. By introducing events that potentially take away progress points, players that ignore security may find themselves losing progress points before they're able to finish their project.

4.6 Initial Feedback

With the design decisions made prior to designing the game, game creation proceeded quickly. There were a total of five working sessions before the game was ready to present. The first four sessions were restricted to the authors. Each working session took approximately one hour and resulted in four to eight hours of adjustment work, particularly in point value adjustments and variable reduction, prior to the next working session. During the fifth session, the game was introduced to a fourth individual who acted as the instructor during the presentation to a collection of instructors. This enabled us to test how quickly someone could learn the game and then teach it to others.

The presentation to the instructors took one hour. During that time, the "instructor" began by discussing the components and then discussed how to play. The instructor "students" immediately began to apply the game mechanics to their class. Prior to beginning play, they were already pointing out aspects of the game mechanics that applied to lessons for their courses. For instance, when the hand size limitation was discussed, one piped up that mirrored the "spend it or lose it" philosophy of acquisitions.

The conversation continued when the instructors began playing the game. During the first two rounds, there was a lot of side conversation about how to use the game in their classes and which game mechanics demonstrated course objectives. During rounds 3 and 4, the players talked less and they focused on the game itself. Then, in round 5, as the game was coming close to a close, the players started talking about game strategy. The game finished in round 7 with a 3-way tie (there were a total of four players). Total play time was about 30 min. The instructors provided valuable feedback, particularly on the need to integrate theme into the components. It was after this demonstration that we began putting specific secure software principles and examples on components.

5 The Experiment

After completing initial development of the DevSecOps game, a pilot study was conducted in a graduate level course on secure software design with 25 students. For this initial pilot study, the game is tested at the end of the course. Seven copies of the game are created with the intention of a mix of three and four player games. The instructor provides an overview of the game, hands it out and has the students play through one or more times. After finishing their playthroughs, the students fill out a survey that measures how well the game taught the learning objectives. Specifically, the survey is designed to test:

- How well does the game force students to judge the cost and risk trade offs between secure and unsecured software aspects including code, data and usability?
- How well does the game provide an environment in which students devise a strategy to produce a system under threats?
- After playing the game, do students comprehend that partial security may equate to no security?
- How well does the game provide examples of insecure code and methods of mitigating it?
- How likely are students to want to play the game on their own?

The survey consists of three questions where the students are able to write answers as well as twenty-seven Likert scale questions taken from [7].

Fokides et al. [7] developed and validated a scale to measure twelve factors in serious games. Each factor has between three and seven questions associated with it. For this research, five factors are relevant and measured by the Likert scale: enjoyment (6 questions), perceived learning effectiveness (6 questions), perceived relevance to (3 questions), perceived ease of use (6 questions), motivation (3 questions). Furthermore, three questions are added to the perceived relevance factor to make six questions with a focus on the game's connection to the course:

- I better understood challenges and concepts presented in the course after playing the game
- The game complemented the course and was a good way to apply the material
- Playing the game prompted additional discussion analyzing course material and its role in the game

The questions taken from [7] are modified slightly. Where the original questions say "use the game", the questions in our survey say "play the game". Furthermore, several of the original questions use phrases like "relevant to my interests" and "things I already know". These are changed to "relevant to the course" and "things in the course". No other changes are made to the original questions. The enjoyment, motivation and ease of use factors are used to assess whether students are likely to voluntarily play the games on their own. The perceived learning effectiveness and perceive relevance factors are used to assess whether students believe the game teaches the learning objectives.

The three long answer questions are:

- One of the learning objectives for the course is identifying security vulnerabilities in software and design specifications.

• What aspects of the game demonstrated the course objective?
• What aspects of the game did not align with the course objective?
• What aspects could be added to better capture the course objective?

- Strategies:

• What strategies did you and the other players follow?
• Which strategies did you find most useful?
• Did those strategies change over the course of the game? if so, why?

- What course concepts did you learn/ have reinforced while playing the game?

The first and third questions are open-ended to see if the student answers demonstrated the game learning objectives. The second question is designed to see what strategies the students discovered and whether they cover the space that the authors intended. Student answers to these questions are coded and used to dig deeper into how effective the game is at teaching the learning objectives, i.e., to assess if students recognize cost and risk tradeoffs, develop strategies, comprehend that partial security may equal no security, identify examples of insecure code and mitigation methods.

After the students complete their play(s), they are given the option to fill out the surveys anonymously and return them.

6 Results

The students self-divided into six groups of four players (with the 25th student watching one of the games). They were then provided with copies of the game and given a 5 min overview and instruction of the game. During that overview, several students took out the game and began reading the rules and looking at the cards themselves. After the instructions were completed, the students each played a single game and then took the survey. The rules explanation took 5 min while play time varied from 25 min (for four groups) to 35 min (for the other two groups).

The primary Likert scale measurement considered whether the game taught the learning objectives. The mean scores for the learning effectiveness factor ranged from 2.79 to 3.92. This seems to indicate some learning effectiveness. Furthermore, the perceived relevance factor had mean scores of 4.08, 3.33, 3.79, 4.21, 3.67, and 2.83. The only low score, 2.83, was for the question "playing the game prompted additional discussion analyzing course material and its role in the game." Unfamiliarity with the game and a neutral score on ease of use may have contributed to a decreased amount of discussion which dropped this question's score. Excluding the discussion question, the results show that students consider the game relevant to the topic. Overall, using Fokides factors, it appears the game is relevant and taught the learning objectives.

The secondary Likert scale measurements considered how likely students are to voluntarily play the game. For the ease of use factor, students were neutral. This suggests that while the game is not super complex, it needs more development. As it is, more than one playthrough is needed to understand the game. The mean scores for the motivation factor are 2.83, 3.58 and 3.63. While these scores are slightly higher than for the ease of use factor, they are close enough to be interpreted in the same way. For the enjoyment factor, five of the six questions had means between 3.54 and 3.91. The one outlier, "I had fun studying with this game" had a mean of 3.13. This suggests the students found the game somewhat enjoyable. It is likely the ease of use factor contributed to this factor scoring slightly lower. If so, more plays will increase the "fun factor". Taking the ease of use, motivation, and enjoyment factors together, the game is difficult enough to learn that students are likely neutral on voluntarily playing the game.

While the factors are a good start, the authors coded the long answers to gain more insight into how well the game teaches the learning objectives. Surveys went through an inter-rater reliability quantitative design analysis. From the coding of the long answers, 88% of the students said they either learned about the need to secure software from unexpected events (game learning objective 3) or that they learned about specific methods for mitigating software vulnerabilities (game learning objective 4) or both. Strategies for the game (game learning objective 2) were spread out among the different students with five taking each one with the strategies of high security or high infrastructure followed by either high security or a rush to finish the software being the most successful. While there were some long answers on balancing cost and risk tradeoffs, the differences in inter-rater coding meant that they could not be quantified. Anecdotally, one student summed it up as, "I would say that secure software development is a series of trade-offs – all security features have a cost and certain events can alter your development significantly."

7 Conclusions and Future Work

This paper presents the development and results from developing the DevSecOps educational game using the GDM framework. The result is a short, highly replayable card game with a small form factor that can be provided to each student as an educational tool. When presented to instructors, the game was well received. When subsequently presented to students at the conclusion of a secure software design course, the students found that it taught the lesson objectives. The GDM framework worked well to create a functional game with minimal effort and it will likely work equally well in other lessons or cyber courses.

Based on the feedback from the students in their long answers as well as the authors' firsthand view of the game being played and to increase inferences from the survey results, there are several areas for future work. First, students will be provided with a video instruction prior to playing the game. The brevity design decision was supposed to result in a game that took 15 to 20 min. Instead, the game took between 25 and 35 min. It is expected that providing the students with a video that they can watch multiple times if desired will result in a first playthrough closer to the goal of 15 to 20 min. This will provide better understanding and enable more class time to play the game. Second, in the demographics section of the survey, students will be asked how long their play(s) took, what grade they think they will receive in the class, which group they were a part of and whether they personally finished 1st, 2nd, 3rd or 4th.

There are also several changes to make to the game. The abstraction design decision allows students to play the game the first time through without reading the descriptions or looking at the cards. A consequence is that there were several comments concerning more domain-specific details in the cards (e.g., a generic Inefficient Production being changed to something more secure software oriented, better text to explain how the card title actually improves security, and more synergies between the different components) to drive home the course objectives better. The trade-off between making the first playthrough easier versus providing more domain specific information for later playthroughs will be re-examined. This also extends to more cards with more different secure software issues on them to give the students more examples. One additional idea is that the cost for security should increase by one for each progress point already achieved in the hopes that this will teach that security is better built in from the beginning and more expensive if bolted on later in the process.

References

1. Abt, C.: Serious Games. Viking Compass, New York (1975)
2. Arnab, S., Lim, T., Carvalho, M.B., Bellotti, F., De Freitas, S., Louchart, S., Suttie, N., Berta, R., De Gloria, A.: Mapping learning and game mechanics for serious games analysis. Br. J. Educ. Technol. **46**(2), 391–411 (2015)
3. Breuer, J.S., Bente, G.: Why so serious? on the relation of serious games and learning. Eludamos. J. Comput. Game Cult. **4**(1), 7–24 (2010)
4. Denning, T., Lerner, A., Shostack, A., Kohno, T.: Control-alt-hack: the design and evaluation of a card game for computer security awareness and education. In: Proceedings of the 2013 ACM SIGSAC conference on Computer & communications security, pp. 915–928. ACM (2013)
5. Echeverría, A., García-Campo, C., Nussbaum, M., Gil, F., Villalta, M., Améstica, M., Echeverría, S.: A framework for the design and integration of collaborative classroom games. Comput. Educ. **57**(1), 1127–1136 (2011)
6. Farber, M.: Gamify your classroom. Educ. Digest **81**(5), 37 (2016)
7. Fokides, E., Atsikpasi, P., Kaimara, P., Deliyannis, I.: Let players evaluate serious games, design and validation of the serious games evaluation scale. ICGA J. **41**(3), 116–137 (2019)
8. Gee, J.P.: Good Video Games + Good Learning. Peter Lang, New York (2007)
9. Gondree, M., Peterson, Z.N.: Valuing security by getting [d0x3d!]: experiences with a network security board game. In: Presented as part of the 6th workshop on cyber security experimentation and test (2013)
10. Howard, M., LeBlanc, D., Viega, J.: 24 Deadly Sins of Software Security: Programming Flaws and How to Fix Them, 1st edn. McGraw-Hill Inc, New York (2010)
11. Hunicke, R., LeBlanc, M., Zubek, R.: MDA: a formal approach to game design and game research. In: Proceedings of the AAAI Workshop on Challenges in Game AI, vol. 4, p. 1722 (2004)
12. Irvine, C.E., Thompson, M.F., Allen, K.: Cyberciege: gaming for information assurance. IEEE Secur. Priv. **3**(3), 61–64 (2005)
13. Järvinen, A.: Games Without Frontiers: Theories and Methods for Game Studies and Design. Tampere University Press, Tampere (2008)
14. Jordan, C., Knapp, M., Mitchell, D., Claypool, M., Fisler, K.: Countermeasures: a game for teaching computer security. In: 2011 10th Annual Workshop on Network and Systems Support for Games, pp. 1–6. IEEE (2011)
15. Krathwohl, D.R.: A revision of bloom's taxonomy: an overview. Theory into practice **41**(4), 212–218 (2002)
16. Lehmann, T.: Race for the galaxy (2019). http://riograndegames.com/games.html?id=240. Accessed Dec 2019
17. Long, D.T., Mulch, C.M.: Interactive wargaming cyberwar: 2025. Technical report, Naval Postgraduate School Monterey United States (2017)
18. McGraw, G.: Software security. IEEE Secur. Priv. **2**(2), 80–83 (2004)
19. Michael, D.R., Chen, S.L.: Serious games: Games that educate, train, and inform. Muska & Lipman/Premier-Trade (2005)
20. Myrbakken, H., Colomo-Palacios, R.: Devsecops: a multivocal literature review. In: International Conference on Software Process Improvement and Capability Determination, pp. 17–29. Springer (2017)
21. Okolica, J.S., Lin, A.C., Peterson, G.L.: Devsecops (2020). https://boardgamegeek.com/boardgame/307111/devsecops

22. Pendleton, A., Okolica, J.: Creating Serious Games with the Game Design Matrix, pp. 530–539. Springer (2019). https://doi.org/10.1007/978-3-030-34350-7_51
23. Shostack, A.: Elevation of privilege: Drawing developers into threat modeling. In: 2014 {USENIX} Summit on Gaming, Games, and Gamification in Security Education (3GSE 14) (2014)
24. Sicart, M.: Defining game mechanics. Game Studies **8**(2), n (2008)
25. Walk, W., Görlich, D., Barrett, M.: Design, dynamics, experience (DDE): an advancement of the MDA framework for game design. In: Game Dynamics, pp. 27–45. Springer (2017)

A Single-Board Computing Constellation Supporting Integration of Hands-On Cybersecurity Laboratories into Operating Systems Courses

Jason Winningham(✉), David Coe, Jeffrey Kulick, Aleksandar Milenkovic, and Letha Etzkorn

University of Alabama in Huntsville, Huntsville, AL 35899, USA
{winninj,coed,kulickj,milenka,etzkorl}@uah.edu

Abstract. Integration of relevant hands-on cybersecurity content into laboratories supporting our required operating systems course helps to provide all computing students a baseline level of cybersecurity knowledge, even if those students never enroll in a dedicated cybersecurity course. However, these hands-on laboratories often require rebuilding of kernels, capturing network traffic, running offensive cybersecurity tools, so that they either require dedicated computer infrastructure, thus increasing the costs in budget-strapped higher education institutions or frequent administrative actions in case of shared infrastructure. To address this problem, we have developed a low-cost single-board computing constellation and support infrastructure as a cost-effective means of delivering both instruction in traditional operating systems topics and related cybersecurity concepts, such as two-factor authentication and isolation via virtualization.

Use of physical infrastructure instead of virtualized laboratory infrastructure has several advantages. The computing constellation offers simplified configuration and management, flexibility, isolation, deployment of diverse laboratory assignments, and uniformity of user experience. The use of physical infrastructure built upon a low-cost single board computer makes the computing constellation scalable to address rising enrollments.

This paper includes an overview of the security-focused laboratories added to the operating systems course, and the construction and operation of the computing constellation infrastructure so that other educators may replicate the infrastructure and integrate similar hands-on cybersecurity laboratories into their existing computing curriculum.

Keywords: Cybersecurity education · Operating systems · Computing infrastructure

1 Introduction

The infusion of cybersecurity education throughout the computing curriculum addresses several pressing problems in computing education. How do computing students who do

K.-K. R. Choo et al. (Eds.): NCS 2020, AISC 1271, pp. 78–91, 2021.
https://doi.org/10.1007/978-3-030-58703-1_5

not wish to specialize in cybersecurity gain the knowledge required to engineer security into products during development? And, how do educators get more students interested in tackling the many challenging problems that arise in cybersecurity?

For too long cybersecurity has been viewed as an afterthought that could be addressed after product development by bolting on security afterwards. Given the almost weekly reports of major new cybersecurity vulnerabilities and successful cyber breach and ransomware incidents, it is readily apparent that this approach to cybersecurity is no longer adequate. Moreover, cybersecurity is often treated as a specialty within computing education with the majority of cybersecurity education occurring in dedicated cybersecurity courses which may only be required of computing students specializing in the cybersecurity discipline. This does not address the pressing need for computing students to understand how to design security into products from the outset and how to consider security issues related to product deployment.

The growing need for employees with cybersecurity skills is another pressing problem worldwide with estimates of millions of new cybersecurity job openings that will need to be filled in the coming years [1]. Nationally, the U.S. Bureau of Labor Statistics (US-BLS) forecasts a 32% increase in the demand for Information Security Analysts from 2018 to 2028 [2]. The US-BLS also forecasts increased demand for cybersecurity-related jobs including computer and information systems managers and forensics technicians.

To facilitate the infusion of cybersecurity education throughout the University of Alabama in Huntsville (UAH) computing curriculums, the authors received a grant from the National Security Agency to develop a set of hands-on, reusable cybersecurity learning modules that could be integrated into commonly required computing engineering courses. Hands-on cybersecurity learning modules were developed for required courses on computer programming, data structures, software engineering, networking, operating systems, and embedded systems. The modules include a variety of resources such as slides introducing the cybersecurity topics, sample solved problems, and step-by-step laboratory manuals. These NSA-sponsored learning modules are being distributed online (https://sites.google.com/uah.edu/cybermodules) so that other instructors may also integrate hands-on cybersecurity laboratories into their own courses at low cost [3].

When integrated into the targeted required computing courses, this set of modules gives all computer engineering undergraduate students exposure to cybersecurity concepts and practical cybersecurity experience via hands-on laboratory exercises. Now all UAH computer engineering students gain a baseline level of cybersecurity knowledge, even if they never take a dedicated course on cybersecurity. With all students in the computer engineering curriculum actively engaged in learning about cybersecurity, the authors hope that more students will choose to focus their studies on cybersecurity, which would help to meet the national demand for more professionals trained in cybersecurity.

Over the years the authors have experimented with several different approaches for delivery of hands-on laboratory experiences in our computing and cybersecurity courses. Each of these approaches addressed different aspects of our resource and facilities constraints. As a representative example, we selected the operating system course and its associated cybersecurity modules since they illustrate a variety of constraints that other instructors may also experience when they try to integrate hands-on cybersecurity laboratories into traditional computing courses.

In subsequent sections of this paper, the authors present the challenges of updating the operating systems (OS) course with new cybersecurity modules, the updated set of hands-on laboratories, deployment options for the cybersecurity laboratories, the architecture and construction of the single-board computing constellation, and our experiences using the computing constellation to deliver the hands-on cybersecurity laboratories. Lastly, the paper discusses lessons learned as a result of using the computing constellation for deployment of the hands-on, cybersecurity learning modules in the operating systems course.

2 The Challenges of Updating the OS Course

One of the challenges faced in the development of the cybersecurity learning modules was the need to find a way to add cybersecurity content to existing courses which were already packed with content. Any solution also had to satisfy resource constraints with respect to facilities, scalability, and accessibility to online students. Discussed below are both the resource constraints and the content constraints.

2.1 Resource Constraints

The integration of cybersecurity laboratories had to satisfy constraints on our existing computer lab facilities but also be scalable and accessible to online students. With current enrollments, multiple sections of the operating systems laboratories are taught in the same 35-seat Linux computer lab. This same physical lab is shared amongst multiple courses including introductory programming, data structures, and software engineering. Throughout the day, sections of these other courses are interspersed with sections of the operating systems laboratories. Since the Linux systems must remain usable for other courses at all times, we are not allowed to give the operating systems students administrative access to these machines to install software, change network settings, monitor network traffic, etc. that may be required for delivery of the cybersecurity laboratory content.

Any deployment of the cybersecurity laboratories must also be scalable and accessible to online students. In the past we have coped with scalability by adding additional in-person laboratory sections, which is governed ultimately by our ability to support additional laboratory sections. We are seeing increased interest from students wishing to register for online sections of courses. A laboratory deployment that supports online access will help us address long-term scalability issues related to access to the physical laboratory environment.

2.2 Content Constraints

The melding of cybersecurity with the operating systems course proved to be particularly challenging with respect to the need to continue to address traditional topics with operating systems laboratory assignments. Prior to this activity the undergraduate operating systems course covered traditional aspects of operating systems in both lecture and laboratory exercises. Cybersecurity topics such as authentication, virtualization, and encryption were covered during lecture, but the laboratories themselves did not provide the student hands-on cybersecurity experience. In addition to weekly laboratory assignments, students would also complete a 4-week end of term project on distributed computing of complex mathematical problems like computing the Fast-Fourier Transform or sorting large data bases, over all 35 lab computers.

To infuse cybersecurity into the operating systems course, our decision was to replace the 4-week project with a series of discrete cybersecurity laboratories utilizing the computing constellation. Our rationale for replacing the hands-on project with cybersecurity-focused laboratories was as follows. We have an existing senior elective in parallel computing which covered this material in substantially more depth, and we also have a required undergraduate course in networking which covered the foundations on socket communications.

3 New Operating System Cybersecurity Laboratories

Presented below is a summary of the new hands-on laboratory experience for students enrolled in our operating systems course. Note that several of the exercises require administrator access for installation of software, capture of network traffic, etc. Moreover, the installation and configuration process required for some labs is lengthy, and thus it may span multiple scheduled lab periods so students need to be able to start, suspend, and resume work without interference over the span of multiple days.

Use of AppArmor Access Control Software. The goal of this laboratory is to demonstrate how an access control software can control the capabilities of a given piece of software through a permissions file.

Lab Contents: The students install AppArmor in a Linux kernel. They then exercise an application in training mode under control of the AppArmor profile generator which generates a description of the system support required by the application. Following that they then run the application under the constraints of the profile they generated earlier.

Use of 2-Factor Authentication to Impede Brute-Force Login Attacks. The goal of this laboratory was to demonstrate for the students how attacks using a port scanner to detect open ports and password crackers can allow access to systems only protected by userids and passwords. Another goal was to demonstrate how the use of Google's two factor authentication capability thwarted these attack mechanisms.

Lab Contents: The students stand up two Linux virtual machines (VMs), one the attack VM and one the target VM on a private network. On the target VM they set up a user account with an easily guessed password. On the attack VM they then launch an **nmap** port scan against the target revealing an open telnet port. Following this they run the **hydra** password cracker against the target allowing access to the target through the easily guessed password. The students then install the Google two-factor-authenticator module on the target VM and the code generator on their cell phones. They then demonstrate that the hydra attack is now thwarted. They also explore the issue of time synchronization by forcing the time on the target machine to differ from that on the cell phone thus disabling the keys generated by the code generator.

Use of the Wireshark Network Scanner. The goal of this laboratory is to demonstrate the power of state-of-the-art network scanners by collecting network traffic and identifying the attacker and target protocol being attacked.

Lab Contents: The students set up a Wireshark network scanner and collect data from a set of managed traffic. They then use the Wireshark scan tool to analyze the network traffic. They decrypt an encrypted packet using a decryption key that we supply.

Use of Containers to Permit Distribution of Software with All Dependencies Resolved. The goal of this laboratory is to demonstrate to the students how the use of containers provides a low cost (in comparison to VMs) vehicles for deploying dependency free software distributions.

Lab Contents: The students try to run several different Python scripts that require different versions of Python. After understanding the special world of dependency conflicts, they stand up a container which has exactly the version of Python that is needed for the distributed application.

Use of QEMU to Emulate a Board Before Construction. The goal of this laboratory is to demonstrate to students that the same emulation technology that has been used for virtualization can also be used to emulate boards before development or by an attacker who does not have physical access to a hardware platform being targeted.

Lab Contents: The students stand up a QEMU emulation of a board with an ARM processor that differs from the ARM processor on the host. The board has a processor, memory, and IO interface. They compile C programs using a different C compiler than that used for the host. They then run these programs on the emulated board. They emulate the board to the level of fidelity of inserting an input on a serial interface and trigger execution of an interrupt handler they have written.

4 Cybersecurity Laboratory Deployment Options

4.1 Laboratory Deployment Alternatives

Over the years, the authors have experimented with a variety of techniques for deployment of hands-on laboratories for computing and cybersecurity content in operating systems and dedicated cybersecurity courses. Prior to pursuing construction of the single board computer (SBC) constellation, the authors considered the following options. Below is a summary of the options considered along with their advantages and disadvantages.

Physical Laboratory Desktop Workstations. With this option students were given administrative access to Dell workstations housed in a small computer lab. With administrative access, students were able to install and configure all software required for the laboratory experiments such as operating systems, hypervisors, and network traffic capture tools. Each student required on-going access to their assigned computer for the duration of the term. The primary advantage of this approach is that students could readily accomplish all of their laboratory learning objectives without interference from other students. The scalability of this approach was problematic since each student added to the course required an additional network drop, an additional workstation, and laboratory furniture in a room that is already at capacity. During the semester, the expense of this physical solution could not be amortized across multiple courses utilizing these same facilities during the same term since each student needed exclusive access to a machine during the semester. These individual workstations also lack battery backup so when power glitches appear, a student may lose a significant amount of work.

Fully Virtualized Deployment for Use on Student-Owned Computers. With this approach, students are permitted to download a set of preconfigured VM targets, such as Metasploitable, along with an attacker VM such as Kali Linux. A clear advantage of this approach is lower cost to the university. An important disadvantage with this approach is that it is hard to maintain identical experimental environments when students launch VMs on their own personal computers. Although the UAH College of Engineering requires students to purchase laptops that meet a minimum specification, in practice students show up with a wide variety of laptop configurations. Many of these configurations are incapable of running the larger problems that require multiple VMs. Network configuration variations are also an important issue since the students' local routers often have behavior different than what is desired for our experiments, such as having dynamic host control protocol (DHCP) service which provides IP addresses at will.

Fully Virtualized Deployment Hosted on University-Owned Computers. With this approach, a campus-owned virtualization server hosts all virtual machines utilized for the laboratory assignments. Each student is provided access to their own private network with a Kali Linux attacker VM and a set of VM targets. A significant advantage of this approach is that with the right virtualization software, scripting may be used to automate stand up and take down of different combinations of VM targets throughout the semester. This allows the instructor to readily reconfigure the environment for the next assignment by switching to an alternative set of VM images. This approach also facilitates remote access for all students including online students since students use SSH to remote into the server before accessing the virtual console of their Kali VM. With enough computing and memory resources, the server is capable of being shared simultaneously among multiple courses. One disadvantage of this approach is that the boot time of the laboratory VMs increases as the number of VMs increases. Another disadvantage of this approach is both the initial capital outlay to purchase the virtualization server along with its supporting infrastructure such as battery backup system and upgrade costs to increase resources to accommodate additional students.

Single-Board Computers Deployed Along with Physical Computer Workstations. With this approach, students utilize university-owned workstations to configure and program the single-board computers. The instructors use this approach exclusively for certain laboratory content such as embedded systems. An advantage of this approach is that each student has access to everything they need at their assigned station. Additionally, the low cost of the boards (~$60 without accessories) makes it affordable for students to purchase their own SBC and complete their assignments remotely. An individual SBC can be readily reconfigured for different lab assignments by swapping out the disk image used for booting the device. This approach suffers from scalability problems with respect to room capacity and the costs of adding new stations for each new student enrolled. The physical access to each workstation also leads to variations in the lab station configurations over time which makes it more difficult for instructors to quickly diagnose problems that arise.

4.2 Single-Board Computer Constellation

The solution we adopted, the SBC constellation, is a set of individual SBCs with managed network infrastructure and server support. It combines the advantages of the SBC solution (low-cost, easy reconfiguration) with the advantages of managed centralized server infrastructure housed in a controlled server room environment (ease of reconfiguration, reliability) and allows us to meet all the requirements of our academic activities in a cost-effective way.

Furthermore, one goal of this funded effort was to develop reusable modules that could be duplicated across other organizations that might have limited system management staff or hardware resources. The SBC approach provides a low-cost mechanism for replication of our environment independent of the starting infrastructure of the destination organization. High schools, junior colleges and 4-year universities can replicate this configuration at a modest cost of about $100 per station.

Finally, we have considered online students, which our university is encouraging our department to recruit. The configuration we have developed will allow students both local to our campus and remote from our campus to have the exact same computing experience since they all access our resources the same way. Because the resources are so modest it may be possible for remote students to build their own laboratory equipment that provides many similar experiences. However, things like measuring network traffic impacts of varying levels of virtualization and containerization will not be able to be exactly duplicated by remote students hosting their own hardware.

5 Constellation Requirements, Architecture, and Operation

To deliver the set of cybersecurity laboratories for the operating system course, the authors developed the following set of requirements for the single-board computing (SBC) constellation.

- Each student must be able to work at their own pace.
- Each student must have remote access to the SBC constellation.
- Each student must be able to view the console of their assigned SBC.
- Each student must have root/administrator access to their assigned SBC.
- Each SBC must be easily reset to a known starting configuration.
- Each SBC must have external network access so that any required software may be downloaded.
- Any cyberattack tools running on the SBC constellation must be isolated from the campus network.
- Each student should be able to view the serial console to allow viewing of boot error messages.
- Each student should be able to remotely power cycle their SBC.
- The SBC constellation should be expandable at relatively low cost to support increased enrollment.

Discussed below is our selection of the SBC computer, the system architecture of the SBC constellation, and operational issues from the student and instructor perspectives.

5.1 Single-Board Computer Selection

For a previous grant from the National Security Agency (grant H98230-16-1-0336), the authors investigated the performance of several single board computers to identify computing platforms that might serve as the basis for the creation of a more secure Programmable Logic Controller that employed hypervisor virtualization to enhance isolation of critical tasks. At that time, it was found that the Odroid XU4Q included enough processing power and memory resources to support the Linux operating system with KVM hypervisor virtualization and multiple small virtual machines.

The Odroid XU4Q single-board Linux computer was selected as the basis of the computing constellation as a result of its ability to support hypervisor virtualization and its low cost [4]. The XU4Q features four ARM Cortex A15 2 GHz processor cores and four ARM Cortex A7 processor cores for a total of eight cores. The board also features 2 GB of onboard RAM along with eMMC and microSD connectors for attachment of flash storage. The XU4Q supports a number of interfaces including gigabit ethernet, USB 3.0, USB 2.0, HDMI, GPIO, and serial console.

5.2 The SBC Constellation Architecture and Operation

The system architecture of the SBC constellation, depicted below in Fig. 1, is driven by two key design considerations. The first requirement was to provide each enrolled student access to their own Odroid XU4Q computer for the duration of the course. This allows students to work at their own pace modifying existing kernel configurations and rebuilding kernels to incorporate the modifications without interference from other students' activities. The second requirement is to minimize the workload on our system manager while providing the richest possible experiences to all our students.

Architecture Overview. The SBC constellation shown in Fig. 1 consists of a server acting as a bastion host and providing DHCP, trivial file transfer protocol (TFTP), and network filesystem (NFS) services to the SBCs; sixty Odroid SBCs; a private ethernet network; USB to serial console adapters and USB hubs; and a remote power management system (now under construction). Students access their assigned SBC via the bastion host and download the software required to complete the laboratory assignments. The USB to serial adapters provide students access to the serial console for debugging. The power management system will allow students to cycle power to their assigned SBC.

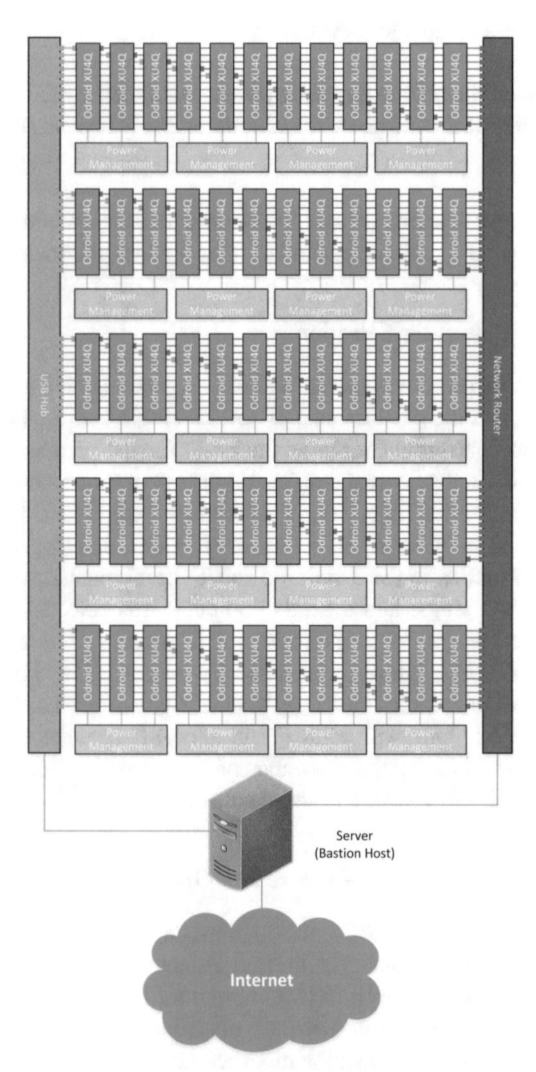

Fig. 1. Block diagram of Odroid XU4Q computing constellation which includes ethernet connections, USB connections for access to serial consoles, and proposed power management interface to allow remote power cycling of individual SBCs.

In its current configuration, the individual SBC computers use PXE to boot an NFS mounted root filesystem. Although the individual student can still damage the files, it is a quick repair to reinstall the files for the individual student because they physically reside on the NFS server. Each SBC has a separate filesystem on the server so that one student's misadventures do not affect any other students. To help students debug boot-time issues, we have connected the serial console via a USB to serial adapter so that a student can see the boot process and can watch their misadventure play out in real time. To further reduce the workload on the system manager, we are installing remote power management for each SBC board so that a student can reset the board on their own if a misadventure crashes their system.

The Odroid community has developed a network boot process based on Intel's Pre-boot Execution Environment (PXE) for the Odroid platforms. It is not a true PXE boot; it instead uses a locally installed PXE bootloader instead of obtaining one from a network server. The network boot process is as follows:

- SBC PXE bootloader broadcasts DHCP request
- DHCP server responds with SBC's IP, netmask, and "next-server" IP
- SBC bootloader issues TFTP request for PXE configuration file
- TFTP server responds with requested file
- SBC bootloader parses the configuration file for kernel and U-Boot device tree binary (DTB) file names
- SBC bootloader requests kernel image file
- TFTP server responds with requested file
- SBC bootloader requests DTB file
- TFTP server responds with requested file
- SBC begins kernel boot process
- SBC kernel performs NFS mount of root filesystem
- Normal system boot continues

Concept of Operation. The bastion server acts as the gatekeeper for the constellation. In order to use our facility students must perform a few tasks. First, they must login to the server (via SSH) which in turn provides SSH, serial console, and power management access to the individual SBCs. They are authenticated to the SBCs separately from the server authentication since the cluster might be used for different laboratories throughout the term (although currently it is dedicated to the operating systems and virtualization courses). After authenticating, the student logs into the SBC via SSH. The student may connect to the console using a terminal application or power cycle the SBC as needed for troubleshooting their work.

5.3 The Odroid XU4Q Computing Constellation

A 60-node Odroid XU4Q constellation was constructed to support the operating systems hands-on learning modules. Figure 3 shows the completed 60-node SBC constellation. For constellation development, we selected the fanless model of the Odroid XU4, which improves overall reliability by elimination of moving parts from the system. Figure 2 below shows a photograph of a single Odroid XU4Q. A bracket was designed and fabricated to allow mounting the Odroid on an industry standard IEC 60715 DIN rail. DIN rail allows for reliable, inexpensive and relatively dense mounting of the SBCs and supporting components such as power supplies, power management, and USB hubs. Each Odroid occupies 1.5″ of DIN rail linearly, and the full array of support hardware occupies a similar average length so that 12 Odroids, their power supplies, power management, and USB console support occupy approximately 3′ (1 m) of DIN rail. We chose to mount this system in an unused 19″ rack, but it could be similarly mounted in a cabinet, on a bench, or on a wall (e.g. in a wiring closet).

Instead of using the vendor-supplied transformer we chose to use industry standard DIN rail mounted industrial power supplies. The Mean Well MDR-60-5 chosen provides 10A at 5 V and is sufficient to supply three SBCs from a single supply. This style of supply is more reliable than an inexpensive consumer supply and simplifies connectivity at the expense of requiring the user to wire the high voltage side of the supply. At the time of construction, the power supplies cost $20 each or $6.67 per SBC.

The serial console connectivity is provided by the Odroid manufacturer's "USB-UART module kit". This kit supplies the USB to serial adapter in PCB form, a short cable to connect the XU4Qs UART port to the USB to serial adapter, and a standard micro-USB cable. A sufficient number of USB ports to connect all 60 devices is provided by DIN mounted 10 port USB hubs with screw terminals for connecting DC power. Cost for serial console connectivity is $25 per SBC.

The original power distribution consisted of cables wired directly into the power supply screw terminals. When the need for power management became clear, design began for a power management system. The principal components are an Atmel ATmega328 microcontroller and VN5E160S high side switch. Connectivity is provided by an RS485 bus. We also included instrumentation in the form of TI's INA219 power monitor. The cost of the current design is estimated to be $12 per SBC.

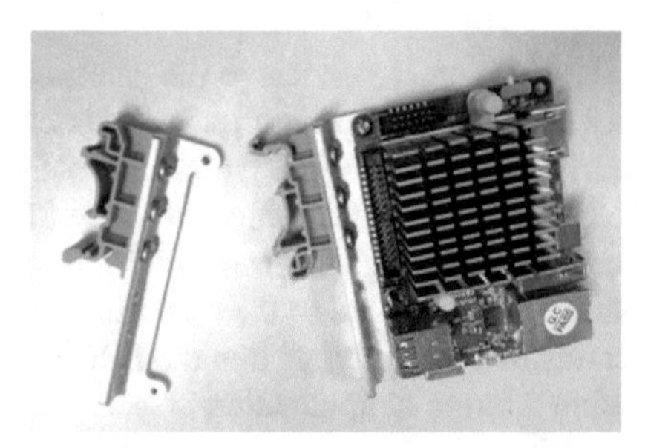

Fig. 2. Photograph of Odroid XU4Q with mounting bracket used to attach an individual circuit board to the mounting rail.

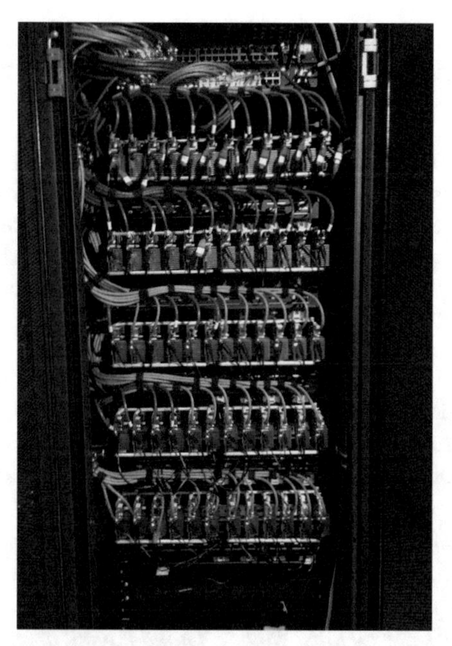

Fig. 3. Photograph of the completed sixty-node Odroid XU4Q constellation in operation. Note that power supplies have been relocated behind the SBCs to increase mounting density.

6 Laboratory Deployment Experiences

Several important lessons were learned from the initial deployment of the new laboratory modules and the Odroid computing constellation in the January 2019 offering of our required operating systems course. The authors have since made improvements to address the issues that arose. Below is a brief overview of the issues and our solutions for those instructors wishing to adopt our lesson materials and deployment approach.

Diagnosing Boot Errors. The installation and configuration procedures associated with the laboratory require students to build the entire system from the ground up. This may include installation of the host operating system, hypervisor, configuration of the associated virtual network, and the creation and booting of several virtual machines. The set of procedures for completing this laboratory are particularly lengthy and as a result some students made errors by skipping or mistyping instructions resulting in systems that were misconfigured. Additionally, many of our students are involved in system support in their day jobs and oftentimes brought their existing mis-information to bear in their course laboratory work. In this case, remote access via the SSH connection proved to be inadequate for debugging purposes since some of the boot process error messages are only visible via a console connection to the XU4Q board itself. Since the Odroid computing constellation is housed in a locked server room, the students who encounter boot-related errors were often unable to view the error messages.

The updated operating systems course which uses the Odroid computing constellation to support the cybersecurity learning modules is now in its second offering as of

January 2020. For the current offering of the operating systems course additional hardware has been purchased. This will allow students to remotely interact with the serial console.

Unbootable Micro SD Memory Cards. Although we originally intended to have the students store their system images on an embedded SD card, it quickly became clear that the students can easily corrupt the SD card, which thus required extra work by the system manager physically removing, re-imaging, and replacing each corrupted SD card. We subsequently implemented network boot capability for these systems which would install a clean disk image through copying files on our file server in a matter of seconds.

Positive Student Feedback on Cybersecurity Labs. The OS course instructor received positive feedback on the new cybersecurity modules delivered via the SBC constellation in the course evaluation surveys. Below are some sample comments:

- *"The Odroid labs were particularly useful, being given administrator access to configure security settings and test security tools out will be great practical experience I believe I will use in future employment."*
- *"Again, enjoyed the security labs. Wouldn't mind having an entire class dedicated to labs like that."*
- *"This class was difficult and one of my least favorite in the CPE curriculum (due to the challenge). Some parts were very interesting, like the security modules."*

7 Conclusions

The infusion of cybersecurity education throughout the computing curriculum ensures that all students become familiar with security threats and understand how to apply sound techniques to mitigate those threats in their hardware and software artifacts. In support of this approach the authors developed a range of teaching modules and hands-on laboratory assignments infused in a range of core computer engineering courses, including computer programming, data structures, software engineering, networking, operating systems, embedded systems, and cloud computing. However, focus on hands-on and experiential learning of cybersecurity modules, coupled with increased enrollments and ever-tightening budgets in higher education, place additional strains on the existing laboratory infrastructure. To address this challenge in an undergraduate operating systems course, the authors developed a single-board computing constellation using inexpensive single board computing platforms.

The advantages of our constellation deployment are as follows:

- The manageability of the SBCs was significantly improved with the constellation approach. The additional hardware cost more than offset the reduction in the management effort during configuration and daily operation.
- The constellation can be readily configured to support other laboratories as well as other courses such as parallel programming.

- By locating all the physical boards on a private network any inappropriate network traffic can be easily isolated from the campus network, thus preventing intentional or inadvertent insider attacks.
- The constellation approach provides reliability and a uniformity of experience not available when students are in full control of the SBC or VM used to perform the exercises.
- Assuming basic computing infrastructure is available, the constellation provides individual computing capability for cybersecurity exercises for 60 students for $6000. Acquiring traditional desktop computers to dedicate to this task would cost $60,000.

Future work includes development of return oriented programming (ROP) attacks and hypervisor attacks. In addition, the constellation can support other courses such as high-performance computing.

Acknowledgments. The authors wish to thank the National Security Agency for supporting this research under grant H98230-17-1-0344. We would also like to thank Mr. Prawar Poudel for his help in developing the laboratory assignments.

References

1. Cybersecurity Talent Crunch to Create 3.5 Million Unfilled Jobs Globally by 2021. Cybercrime Magazine. https://cybersecurityventures.com/jobs/. Accessed 25 Jan 2020
2. Information Security Analysts. U.S. Bureau of Labor Statistics Occupational Outlook Handbook. https://www.bls.gov/ooh/computer-and-information-technology/information-security-analysts.htm. Accessed 25 Jan 2020
3. Coe, D., Kulick, J, Milenkovic, A., Etzkorn, E.: Hands-On Learning Modules for Infusion of Cybersecurity Education Throughout Computing Curricula. https://sites.google.com/uah.edu/cybermodules. Accessed 25 Jan 2020
4. Hardkernel website. https://www.hardkernel.com/shop/odroid-xu4q-special-price/. Accessed 25 Jan 2020

TWOPD: A Novel Approach to Teaching an Introductory Cybersecurity Course

Chola Chhetri(✉)

Northern Virginia Community College, Annandale, VA 22030, USA
cchhetri@nvcc.edu

Abstract. In today's world, technology is ubiquitous. Today's children grow up with a lot of technology, such as phones, gaming systems, tablets and computers. Even 'things' are connected to the Internet. Such things include door locks, coffee makers, refrigerators, thermostats, and speakers to name a few. There are more Internet of Things devices than there are people worldwide. Connected things have changed lives of people and the world in a lot of positive ways, but they also come with numerous security and privacy issues.

Thus, cybersecurity education has become crucial in providing security awareness, secure behavior and cybersecurity expertise. This paper discusses TWOPD, a novel approach to teaching cybersecurity in a college classroom and evaluates the effectiveness of the proposed approach.

Keywords: Cybersecurity · Teaching · Learning · Cybersecurity education · Writing · Projects · Transparent assignment

1 Introduction

We live in an age where the number of personal computing devices has outnumbered the entire population. In 2016, 85% of United States (US) households contained at least one smartphone, desktop or laptop computer, tablet, or streaming device. The "typical/median" household contained five such devices [1]. Adoption of internet-connected technology is on the rise. Gartner predicts there will be over 20 billion Internet of Things (IoT) devices worldwide by 2020 [2]. This number does not include computers and phones.

About 15 billion records containing personal information have been breached in the past five years, effecting almost all sectors of industry, including social media, hospitality, technology, retail, entertainment, non-profit, government, healthcare, education, and finance [3, 4]. Data breaches have large financial and psychological impacts on individuals, and businesses suffer huge financial losses sometimes leading to bankruptcy.

In addition to attacks that effect individuals, such as identity theft, attacks of large scale, such as the Mirai Distributed Denial of Service (DDoS), have already occurred. The Mirai botnet brought down large businesses effecting millions of users [5]. With the rapid increase in adoption of connected technology, we can only expect cybersecurity incidents to rise [2].

K.-K. R. Choo et al. (Eds.): NCS 2020, AISC 1271, pp. 92–99, 2021.
https://doi.org/10.1007/978-3-030-58703-1_6

Thus, there is a growing need for cybersecurity awareness for everyone and cybersecurity education at all academic levels. This has led to an increase in cybersecurity programs at educational institutions, cybersecurity certifications in the industry, and job demand in the workforce.

In our institution, the introductory cybersecurity course, ITN 260, is mapped to CompTIA Security+ certification [6]. It is designed to provide students with fundamental cybersecurity knowledge, while preparing them for higher level cybersecurity courses and entry level industry certifications.

In this paper, I discuss the novel TWOPD pedagogical approach used to teach cybersecurity curriculum in the ITN 260 course (referred to as 'pilot course' in this paper) and present findings of its evaluation. Evaluation shows the approach is effective in terms of student satisfaction level.

The contributions of this paper are as follows: (a) It presents TWOPD, a novel approach of teaching cybersecurity course, and (b) It discusses results of evaluation of the approach.

2 Related Work

Winkelmes (2013) studied transparency in learning and teaching and published a transparent assignment template that contain components: purpose, tasks, and criteria for success [7]. The importance of transparent assignments has been widely studied thereafter, and many instructors have adopted it. Transparent design demonstrates the relevance of the assignment to student learning, clarifies activities, and allows student self-evaluation of the assignment [8].

Writing helps students conceptualize complex ideas, express them, and learn better. Writing-to-learn strategies have been studied in science and other disciplines. Gunel et al. (2007) performed secondary analysis of six studies on writing in science and found that students benefited more from writing-to-learn than science writing and that they scored better [9].

Online courses, when well-designed, have the ability to offer many pedagogical benefits to students, in addition to the flexibility in time and geographical boundaries. The rise of online courses, Massive Open Online Courses (MOOCs), and online degrees is an evidence of this argument. Glance et al. (2003) showed that online courses can be designed to be better and as effective than face-to-face courses [10].

Past research has demonstrated benefits of project-based learning (PBL). Researchers have shown that PBL helps students receive a deeper understanding of concepts as they are engaged in solving problems. Students also perform better than those in traditional formats [11].

Demonstrations are an important method of teaching to humans [12] and robots [13]. Students have various learning styles, and demonstrations provide students an opportunity to visualize the concepts demonstrated and formulate ideas. Demonstrations help students form arguments, questions existing concepts, and generate new research ideas [14].

3 The TWOPD Approach

The approach is derived and enhanced from the best of pedagogical approaches for online and face to face courses. The TWOPD approach comprises of the following elements:

1. **T**ransparent Assignments
2. **W**riting
3. **O**nline Design
4. **P**rojects
5. **D**emonstrations

3.1 Transparent Assignments

A transparent assignment consists of three main components: purpose, tasks, and success criteria [7]. The purpose of the assignment shows a clear connection of the assignment to student learning, learning outcomes, and relevance of the assignment to cybersecurity functions in the real world. When students see the connection of the assignment to their own learning, they are motivated to complete it with greater interest.

The 'tasks' component lists the activities that the student performs. The 'criteria for success' component lists features of a finished assignment and a grading rubric.

All assignments in the pilot course were designed to be transparent. The instructor asked a few colleagues to review sample assignments for transparency before piloting them in the course and revised the assignments further based on student feedback.

3.2 Writing

Writing reinforces learning. Writing allows students to express learned concepts for others to read and learn. However, in the field of information technology (IT), global industry certification exams rely on multiple choice questions and scenario-based or performance-based questions, which have the advantage of automatic grading and instant result notification. Consequently, introducing writing in my classes often faces resistance and is criticized as an added burden.

Writing offers multiple merits, such as helping students learn invisible technology concepts and express those concepts in a clear fashion, which is often overlooked in IT classes. Students who write well have the potential to perform well not only in writing classes but also in other classes, and are likely to move up further in their educational pursuit. In the workforce, good writing skills empower them with the ability to explain IT concepts to coworkers, IT professionals, and people from non-technical backgrounds.

Writing, coupled with communication and presentation skills, is a well-sought skill in the industry. Employers in our region have listed good communication skills as a principal requirement for Information Technology and cybersecurity job positions. Technical professionals, such as cybersecurity engineers, need these skills in order to move up to managerial and leadership positions.

Thus, practicing writing in our courses is preparing our students to thrive and grow in the industry at all levels of jobs. Incorporating writing in the course not only helps develop writing skills in the students but also allows the transfer of learned skills from academia to the workforce [15].

To reap these benefits of writing in learning, the pilot course incorporated writing exercises. Each class meeting, students received writing exercises, which contained questions on the cybersecurity content of previous class meeting. Half of the writing exercises were impromptu. The other half were pre-published in the Learning Management System (LMS). Students were required to have studied the content from last class in preparation for the exercise. Student writings on the exercise questions received instant feedback from peers and instructor.

These writing exercises were designed to help students ensure they have understood the complex cybersecurity concepts and are able to express or share them in their own words. I believe this will prepare them not only to perform cybersecurity functions at work, but also to share and contribute in a team.

3.3 Online Design

The course was designed to be as self-paced and independent as an online course. It was developed using Blackboard, an online LMS. The course began with an orientation to online learning, that provided students familiarity with navigating through the course.

The course content was divided into modules per week. Each weekly module contained the following items:

- Summary of learning activities
- Reading for the week
- Slides from the instructor
- Links to online materials and videos related to the week's content
- Practice quizzes
- Assignments, with instructions and due dates.
- Hands-on projects
- Exams and other assessments
- How to prepare for the face-to-face meeting, and
- Getting ready for the next week

The course was also supplemented with resources, such as books, videos, tutorials, and practice exams, which help in preparation for the CompTIA Security+ certification exam.

3.4 Projects

This course was not only focused on learning concepts by discussion. Projects carried about one-third weight in learning materials and assessments. Three types of projects were incorporated in the course:

1. Short Project Sets
2. Hands-on Projects
3. Long End-of-Semester Project

3.4.1 Short Project Sets

Every project set contained 6–12 short hands-on projects, each with an estimated completion time of about an 0.5–1 h. Students performed two to three graded project sets during the semester. Step by step instructions were provided for each project in the project set. Students submitted screenshots as evidence of work and lessons learnt from each project. Instructor provided clear grading rubric/criteria in the project set instructions.

Instructor provided help and guidance in group and individually, and encouraged students to notify when they encountered problems. Early-start advice was provided via classroom announcement, LMS notifications and email notifications.

3.4.2 Hands-On Projects

Weekly hands-on projects were designed to help students understand the cybersecurity concepts taught every week and were not graded. Students performed about 4–6 hands-on exercises every week and instructor provided help, support and explanation when sought by students.

3.4.3 Long End of Semester (EOS) Project

The instructor published a list of long cybersecurity hands-on projects, with estimated completion times of 10–12 h. Student teams of no more than 2 students picked an EOS project from the provided list of projects. The LMS system was used to track the teams and the project picks to avoid duplication.

EOS projects were graded and grading criteria was published along with instructions. Teams submitted project reports containing rationale, screenshots/evidence of work, and lessons learnt. Teams also performed in-class presentations to share the work and lessons learnt with peers.

3.5 Demonstrations

Demonstrations were found to be useful to help students visualize cybersecurity concepts. They helped build hands-on skills, communication skills, and team skills, which are all key to any cybersecurity job function. In-class demonstrations were performed by both instructors and students.

3.5.1 Instructor Demonstrations

Every classroom meeting included 5–10 demonstrations of cybersecurity concepts. The demonstrations were built in the face-to-face classroom meeting, in such a way that they would provide break from lecture and spark interest in the discussed topics.

3.5.2 Student Demonstrations

Every student was required to perform a demo of one cybersecurity concept in class. Students were allowed to pick a topic of their choice for the demo and the instructor kept track of the demonstrations for grade. In addition to these individual demonstrations, student teams performed demonstrations of their EOS project from Sect. 3.4.3.

3.6 Additional Support

In addition to the five elements of the TWOPD approach, learning was complemented with the following additional support to students in the course:

a) face to face group meetings, that incorporated lectures, review of difficult concepts, scenario discussions, round robin question–answer sessions, quizzes and hands-on exercises, and
b) mandatory individual meetings, that incorporated clarification of difficult cybersecurity concepts, explanation of student-requested material, and help on hands-on exercises and projects.

These sessions were coordinated and conducted by the instructor at the request of students.

4 Evaluation

A survey was conducted at the end of the semester to evaluate the effectiveness of the course. The survey contained 18 questions. A total of 17 questions measured the positive, negative or neutral responses on a Likert scale of 5 (Strongly Agree, Agree, Neutral, Disagree, Strongly Disagree). There was 1 open-ended question that sought suggestions for improvement in the course.

A total of 22 students took part in the voluntary survey. All students entered the course with familiarity of computers and related technology (Agree: 4, Strongly Agree: 18). While this was expected, an orientation to LMS and course navigation was provided. Most students were satisfied with the self-paced training provided (Agree: 4, Strongly Agree: 18). Majority of students were happy with the LMS used (Disagree: 4, Agree: 6, Strongly Agree: 12).

Student response to the suggested approach was largely positive. Majority of students thought that the use of the LMS in the course and the course design helped improve their experience (Neutral: 4, Agree: 2, Strongly Agree: 16). Most students also thought that the assessments tied to the LMS helped them perform better (Disagree: 2, Neutral: 2, Agree: 8, Strongly Agree: 10).

Overall, students seemed largely satisfied with the course (Disagree: 2, Neutral: 4, Agree: 6, Strongly Agree: 10). Figure 1 shows the graphical representation of the overall satisfaction results.

The survey provided students an option to write suggestions for improving the course. One student suggested an improvement of giving an "online quiz every week for a grade to help ease up the load of the face to face" quizzes and their grade distribution. Students liked the mandatory meetings as they provided an opportunity to receive "adequate" time to ask questions. One student suggested that students may benefit even more from additional hands-on exercises in the course.

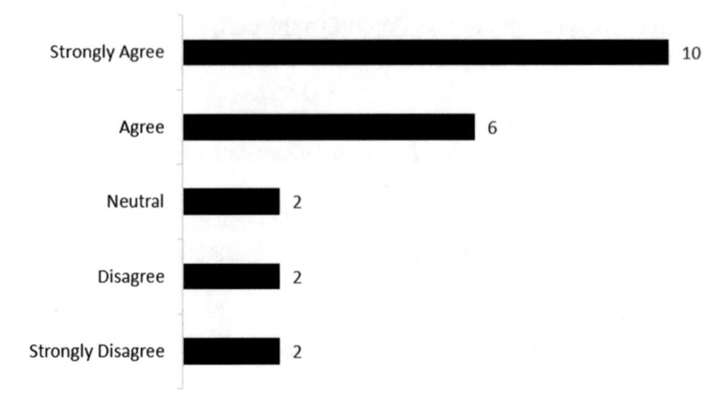

Fig. 1. Most students were highly satisfied with the course.

5 Discussion and Conclusions

The TWOPD approach utilized the best of the face to face and online teaching methodologies. Transparent assignment design allowed clarity in instructor-student expectations, minimizing misunderstood instructions. Students initially were opposed to the idea of writing in a cybersecurity course, citing certification exams not containing writing questions. However, after a few modules the rejection faded, and students began to appreciate the help writing provided in their learning.

Online design of the course allowed students access a full, self-paced, course with learning materials and assignments always at their fingertips. Projects taught students how to perform cybersecurity tasks. Demonstrations allowed them to learn to share their work and communicate in teams. Overall, the course proved to be beneficial to students and students reported a very high level of satisfaction in the course.

The TWOPD approach was effective in achieving high level of student satisfaction in the pilot course and improving student learning experience.

One limitation of the study is its small sample size. The study could be expanded to include more survey results. Another limitation is in the pedagogical approach. It would be desirable to grade all the hands-on projects if resources, such as time and availability of teaching assistants, permit. However, for the study only Project Sets were graded, by carefully choosing the ones that meet the course learning outcomes.

Future work should look into utilizing the approach in other cybersecurity courses and evaluating it at a larger scale. It will also be beneficial to examine whether TWOPD approach could work effectively in teaching hands-on cybersecurity courses, such as penetration testing, where students require a great deal of time to learn the tools themselves.

Acknowledgments. I thank anonymous reviewers for their feedback and the survey participants for their responses. I also thank Writing Across Curriculum 2019 leaders for their support in creating transparent assignments and faculty fellows for their invaluable feedback.

References

1. A third of Americans live in a household with three or more smartphones. Pew Research Center. https://pewrsr.ch/2r0o6he. Accessed 19 July 2019
2. Hung, M.: Leading the IoT. Gartner, Inc. (2017)
3. Breach Level Index. https://breachlevelindex.com/. Accessed 19 Jul 2019
4. Privacy Rights Clearinghouse. https://www.privacyrights.org/. Accessed 19 Jul 2019
5. Antonakakis, M., April, T., Bailey, M., Bursztein, E., Cochran, J., Durumeric, Z., Alex Halderman, J., Menscher, D., Seaman, C., Sullivan, N., Thomas, K., Zhou, Y.: Understanding the Mirai botnet. In: Proceedings of the 26th USENIX Security Symposium, Vancouver, BC, Canada, pp. 1093–1110 (2017)
6. CompTIA Security+ Certification. https://certification.comptia.org/certifications/security. Accessed 20 Jul 2019
7. Winkelmes, M.-A.: Transparent assignment template. Transparency in Teaching (2013). https://tilthighered.com/assets/pdffiles/Transparent Assignment Template.pdf
8. Winkelmes, M.A., Bernacki, M., Butler, J., Zochowski, M., Golanics, J., Weavil, K.H.: A Teaching Intervention that Increases Underserved College Students' Success. Peer Rev. **18**, 31–36 (2016)
9. Gunel, M., Hand, B., Prain, V.: Writing for learning in science: a secondary analysis of six studies. Int. J. Sci. Math. Educ. **5**, 615–637 (2007). https://doi.org/10.1007/s10763-007-9082-y
10. Glance, D., Forsey, M., Riley, M.: The pedagogical foundations of massive open online courses. First Monday **18** (2013). https://doi.org/10.5210/fm.v18i5.4350
11. Kokotsaki, D., Menzies, V., Wiggins, A.: Project-based learning: a review of the literature. Improv. Sch. **19**, 267–277 (2016)
12. Kirk, J., Mininger, A., Laird, J.: Learning task goals interactively with visual demonstrations. Biol. Inspired Cogn. Archit. **18**, 1–8 (2016)
13. Jackson, A., Northcutt, B.D., Sukthankar, G.: The benefits of immersive demonstrations for teaching robots. In: 2019 14th ACM/IEEE International Conference on Human-Robot Interaction (HRI), pp. 326–334 (2019)
14. Cypher, A.: Watch What I Do Programming by Demonstration. MIT Press, Cambridge (1993)
15. Adler-Kassner, L., Wardle, E.: Naming What We Know. Utah State University Press, Boulder (2016)

Network System's Vulnerability Quantification Using Multi-layered Fuzzy Logic Based on GQM

Mohammad Shojaeshafiei[1(✉)], Letha Etzkorn[1], and Michael Anderson[2]

[1] Department of Computer Science, University of Alabama in Huntsville, Huntsville, USA
{Ms0083,etzkorl}@uah.edu

[2] Department of Civil and Environmental Engineering, University of Alabama in Huntsville, Huntsville, USA
andersmd@uah.edu

Abstract. Network security is one of the crucial components of an organization's security system. Much research has been conducted to come up with a clear-cut approach in order to quantify organizations' network system vulnerabilities. Many security standards such as NIST SP-800 and ISO 27001 with the guidelines and clauses are published with a reasonable outline to pave the ground for a safe track towards secure system design, however, these standards do not clearly show the details of work implementation. In this paper, we apply Fuzzy Logic methodology to quantify each factor and sub-factors derived using Goal Question Metrics in network security. Our procedure follows a bottom-up hierarchy model from the details of a security component to the desired goal in order to address vulnerabilities in a quantified manner in the Department of Transportation. Thus, our approach measures different types of potential vulnerabilities in a network.

Keywords: Cybersecurity · Fuzzy Logic (FL) · Goal Question Metrics · Vulnerability · Network security

1 Introduction

There are numerous vulnerabilities affecting organizations' assets. The number and types of vulnerabilities may vary from day to day or from time to time. So the process of vulnerability analysis depends on how and what type of methodology the analysts apply for vulnerability assessment. There are several methods and techniques for threat and vulnerability estimation. They are mainly categorized into two different directions: Quantitative or Qualitative techniques [1]. The qualitative approach is concerned with discovering the probability of an event occurring and the consequences and impact of such an event or malicious attack in the system. In contrast, the Quantitative method uses mathematical and statistical tools to evaluate risk [2]. The purpose of this paper is to use a multi-layered Fuzzy Logic (MFL) to quantify potential vulnerabilities of the organization's network. Our proof of concept will be the Department of Transportation (DOT). Considering all crucial and sensitive components of DOT, we design an integrated and comprehensive model based on all major aspects of service availability, integrity,

K.-K. R. Choo et al. (Eds.): NCS 2020, AISC 1271, pp. 100–115, 2021.
https://doi.org/10.1007/978-3-030-58703-1_7

confidentiality, and accuracy. In this way, we primarily investigate to what extent a potential vulnerability measurement would bolster the security of a network. We apply the Goal Question Metrics (GQM) [3] approach to determine the different perspectives of DOT's network system security that may be susceptible to any type of threat or vulnerability. Considering Availability, Integrity, Accuracy, and Confidentiality as the main factors of Network Security in DOT, we map every factor and related sub-factors to NIST and ISO 27001 security standards, then we build up the security structure for vulnerability quantification based on the appropriate security standards. This procedure will be explained more in Sect. 3.

In this paper, our approach is initially proposed in a qualitative manner to be processed as an input of MFL, in which inputs are derived from an expert evaluation of the real-world security environment of the DOT. Experts' input provides data for fuzzification then we obtain the crisp number which is quantified from the defuzzification process in the Centroid model [4] based on the assigned rules plus membership function (MF).

The rest of this paper is organized as follows: Sect. 2 presents the background of the work which discusses the Fuzzy Logic in more detail, network security measurements and network security vulnerabilities. Section 3 provides a description of our methodology in GQM, security standards, factors and sub-factors of network security and graph model mapped to Fuzzy Logic (FL). Sections 4 and 5 address the implementation processes and conclusion respectively.

2 Background

2.1 Fuzzy Logic

Fuzzy Logic was introduced by Lotfi Zadeh [5] at the University of California in Berkley. In fact, it is defined as a multivalued logic with values between conventional values and is an alternative to traditional notions of set membership. " With FL we should be able to compute linguistic variables, that is, variables whose values are not numbers but words or sentences in a natural or artificial language" [6]. A linguistic variable is a collection of a variable name, a set of terms, universal discourse, a set of syntax rules and a set of semantic rules. The main idea in FL is to convey human language which is ambiguous in essence. For example, the temperature is a linguistic variable that we use in human language and can be a set of terms very cold, cold, warm, hot, etc. With the application of Fuzzy Logic, we should be able to convert such language to numbers. A classical set is the set of variables x such that the variable comes from the universal discourse X and x can be anything such as temperature. $A = \{x\,|X \; and \; X \; is \; a \; form \; of \; temperature\}$; x denoted as temperature (x) and X denoted as hot temperature (X) [5]. Thus, in a fuzzy set A_{fuzzy}, we have Membership Function μ_A: $X \rightarrow [0, 1]$, where μ_A denotes MF and each element of X is mapped to a value between 0 and 1 which is called a Membership value and quantifies the grade of membership of the element in X to the fuzzy set A. In other words, MF quantifies to what degree x belongs to set A since A is a linguistic variable [5].

$$A_{fuzzy} = \{(x, \mu\_A(x)), x \in X, \mu\_A(x) \in [0, 1]\}$$

In order to address the result of the model graphically in FL, we need to have an MF, wherein the x-axis represents the universe of discourse, and the y-axis represents the degrees of membership in the [0, 1] interval. There are different types of MF such as Triangular MF, Trapezoidal MF, Gaussian MF, Generalized Bell MF, etc. [7], in which the most common one is Triangular MF. (See "Fig. 1").

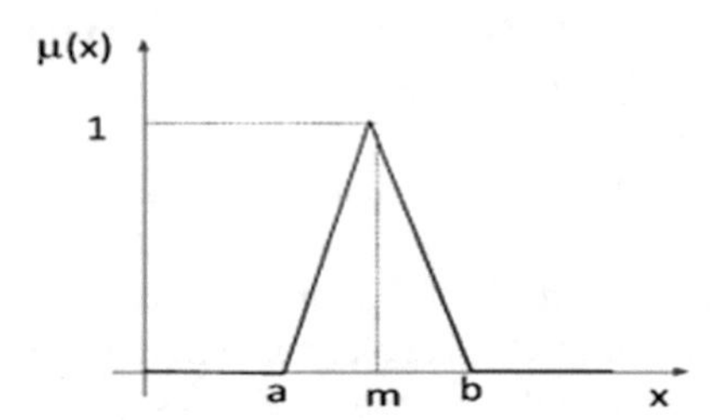

Fig. 1. Triangular membership function

It is defined by a lower bound and an upper bound b and the value m where a < m < b.

$$\mu_A(x) = \begin{cases} 0 & x \le a \\ \frac{x-a}{m-a} & a < x \le m \\ \frac{b-x}{x-m} & m < x \le b \\ 0 & x \ge b \end{cases}$$

The first step of logic processing is a domain transformation called Fuzzification. It decomposes the input into one or more fuzzy sets. In other words, it converts the crisp numerical into fuzzy linguistic quantifiers that involve an imprecise set definition, MF and the degree of actual value representation. This phase determines to what degree MF assigns the input data to each fuzzy set [8].

The second step of the processing is the Inference Engine. Here, all the (if-then) rules and facts extracted from data will be created on the basis of inheritance. The variables are derived from class data which are instances of that class [9].

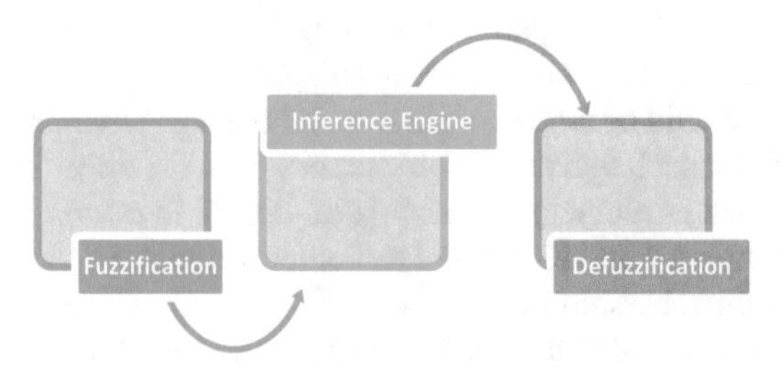

Fig. 2. Fuzzy model

The last step is called Defuzzification which is the process of extracting a quantified value as an output of an aggregated fuzzy set. It transfers fuzzy inference rules into a crisp number [10]. Figure 2 depicts the fuzzy model.

2.2 Related Works

Fuzzy Logic based on the Mamdani-style inference [11] has been applied to identify potential threats in systems and software [12]. The most common models of inference systems are Mamdani Fuzzy, Sugeno Fuzzy and Tsukamoto Fuzzy model [13].

One of the most common approaches to ranking system vulnerabilities is the Delphi approach [14] that prioritize vulnerabilities based on estimation from predetermined metrics in the system. The problem of this method is that the user(s) should rate the priorities over and over to find out the most possible fixed rate. It may be a reasonable method for a small number of assets in the system, but it would be very hard in a large network with numerous assets for evaluation.

Anton et al. [15] introduced the Vulnerability Assessment and Mitigation (VAM) methodology with six-step procedures in a top-down-manner to mitigate any possible threats from the past and to perform current and prediction analysis for the future threats to protect the system against the possible attacks and failures. The authors classified all system components in regard to any cyber and human incidents accordingly to find out which prevention techniques could be the most appropriate detection for vulnerabilities categorization. This method handles the processes of risk reduction with the analysis of vulnerability to match with the related safeguards. It is a reliable approach if we want to consider only threat and risk as important factors of vulnerability measurement.

There has been much research in using fuzzy systems for risk and vulnerability analysis [16–18]. These researchers describe a technique with the assistance of Fuzzy Logic to measure the probability of failure while they measure the likelihood of future attacks in the system. In order to determine potential risk based on linguistic values, Shah [16] applied Key Risk Indicator (KRI), an operational variable that estimates the potential security losses corresponding to the risk in the system (KRI specification). Afterward, they calibrate the fuzzy representation of KRI and the loss amount. For instance, linguistic descriptors (very low, low, medium, high, very high) are assigned to a range of values for each KRI. In the next step, they specify the impact of KRI on the loss amount. For that, fuzzy rules in the form of (if-then) are provided by human experts based on their real-world experience to correspond to different levels of KRI. Table 1 addresses an example of loss amount in software corresponding to the risk based on the complexity and the years of experience in software development.

Table 1. If-then rules for software loss amount [16]

Rule 1:	If	Product complexity	Is	High or
		Years of experience of programming	Is	Low
	Then	Security loss	Is	High
Rule 2:	If	Years of experience of programming	Is	Low and
		Growth rate of software	Is	High or negative
	Then	Security loss	Is	Very high
Rule 3:	…	…	…	…

Zhao et al. [17] proposed a fuzzy risk assessment in network security. It combines the hierarchy structure of the Analytical Hierarchy Process (AHP) method with Fuzzy Logic that is improved according to the actual condition of risk assessment. They estimated the probability of each risk factor based on their degree of importance. The entropy-weight coefficient is used for objective computation.

Lee [18] conducted research on information security risk analysis that combines AHP and FL as a hybrid and a comprehensive method. The author's discussion takes into account that in order to have a precise security evaluation, three fundamental aspects of risk and security are required; 1. Information security analysis, 2. Information security risk assessment and, 3. Information security management. The indexes of criteria are comprised of assets, threats, vulnerabilities and safety measures. In each step, each index brings about new indexes such as confidentiality, integrity, and availability for the new level and accordingly for the next levels.

Watkins et al. [19] proposed a quantifiable approach to assess the cyber maturity level. This method is based on the AHP derived from the importance of security components. It provides quantifiable risk metrics for overall cyber vulnerability measurement and it also prioritizes the severity of the vulnerabilities. They determined the rank of impact, age of vulnerability and exploit availability which is used to determine the impact of vulnerabilities. Hence, the summary of their experience is as follows; First, exploit availability is two times as important as impact. Second, the impact is five times as important as the age of exploit. And finally, exploit availability is seven times as important as the age of exploit. One of the main advantages of this framework is that its results are reproducible and are reliable mathematically.

3 Methodology

3.1 Network Security Standards

ISO/IEC 27001:2013 [20] is an international standard to specify an information security management system (ISMS). This is a systematic approach to determine processing technology that helps to protect and manage the organizations' information based on risk management systems. This standard helps organizations to focus on the three key aspects of information; confidentiality, integrity, and availability. Confidentiality means information should not be available or disclosed to unauthorized users, entities or processes. Integrity means that information should be complete and accurate and also protected from corruption. Availability means that information is accessible and usable as and when authorized users require it [20]. This standard specifies 114 practice controls including physical access control, firewall policies, security staff awareness programs, procedures for monitoring threats, incident management processes and finally encryption policies.

NIST SP800-53 is published by the National Institute of Standards and Technology. It is a catalog of security and privacy controls for federal information systems and organizations in order to "protect organizational operations and assets, individuals, other organizations and the nation form a diverse set of threats including hostile attack, natural disaster, structural failures, human errors, and privacy risks" [21]. It includes the procedures in the Risk Management Framework based on security controls for a federal information system for security requirements. Unfortunately, these standards do not clearly show the details of work implementation to design the system network. They only represent the guideline and required fundamentals of frameworks that organizations need to comply with.

In our research, the very first step is to leverage all components of ISO 27001 and NIST SP800-53 frameworks to meet our network's security requirements. For that, we analyzed DOT's network system to determine the most significant components of security that should be maintained in the network to protect the system against any potential vulnerabilities. Thus the components consist of Confidentiality, Integrity, Availability, and Accuracy (CIAA). We intentionally consider Accuracy as a separate security component in the network security of DOT. These four components are the main security factors of our network framework to evaluate and measure the potential vulnerabilities.

3.2 Goal Question Metrics

In order to articulate our security program in DOT, we apply the Goal Question Metrics approach [3] in the next step of the vulnerability analysis in the system. The main reason for using this methodology is that it is a de facto standard for quality metrics in software engineering [3]. In GQM we analyze the goal(s) we desire to achieve in the project. In our research, the goal is to achieve a quantifiable vulnerability measurement based on the particular set of rules to address data interpretation. The sketch of this part is goal, question and, metrics in a way that fist we define the goal(s) for the object(s) according to the security requirements of the system, question(s) characterize the defined object with respect to security components and, metrics provide measurement procedures in order to answer the questions quantitatively. Measurement goals are defined for the specific needs of an organization and are traceable to a set of quantifiable questions [22]. In our previous research [23], we proposed a hierarchy security model to define a security framework using GQM as an introduction to vulnerability measurement. The main purpose of the research was to find a standard security solution to trace from security requirements to security metrics. For instance, we defined how "media handling" rules should follow to security standards and how based on GQM we can trace it to controls in section A.10.7 in ISO 27001 and clause MP-7 in NIST SP800-53.

3.3 Security Factors and Sub-factors

Based on the main components of network security (that are our main security factors) we designed sub-factors for each factor accordingly. The following pattern in Table 2 shows our definition of security factors and sub-factors for DOT's network security.

As we mentioned before, the goal of this research is to quantify the network system's vulnerability in DOT. Hence, we designed a questionnaire in a top-down manner from GQM. The number of questions for each factor varies from `1` to `n` depending on the context of security analysis. When all questions are designed for the goal(s), we apply the measurement procedure backward with aggregation. For instance, we ask the following questions:

Table 2. Network security factors and sub-factors in DOT

Network security	Availability	Redundancy	
		Monitoring & alerting	
		Backup	
		Load balancing(proxy)	
		Disaster readiness	
		Disaster recovery plan	
		Quality testing	
		Wireless security	
	Accuracy	Data accuracy	Annual review
			VAM (Vulnerability Assessment & Management)
			Data labeling
	Integrity	Data security	Encrypted communication
			Supply chain
		Data quality and integrity	
		Software security	Software update
			Outsource development
			Threat detection update
		Quality testing	
		Wireless security	
	Confidentiality	Authorization and identification	Username and password
			Program access
		Authentication	Data classification
			Two-factor authentication
			Use geographic location as an authentication factor
		User Access Control (UAC)	
		Encryption	
		Quality testing	

Availability

1. Does DOT make sure security mechanisms and redundancies implemented to protect equipment from utility service outages (e.g., power failures, network disruptions, etc.)
2. What percentage of DOT's equipment has security mechanisms and redundancy to protect equipment from utility service outages built-in?
3. Does DOT implement network redundancy in its system?
4. Does DOT perform industry-standard monitoring and alerting on devices that process and store sensitive information?
5. How frequently does DOT back data up?

Integrity

1. How often does DOT ensure that data does not migrate beyond a defined geographical residency?
2. How often are secured and encrypted communication channels used in DOT when migrating physical servers, applications, or data to virtual servers?
3. How often does DOT select and monitor outsourced providers in compliance with laws in the country for data storage, location, processes, and transition?
4. How are third-party software updates (Adobe, Java, etc.) installed on DOT's computers?
5. How often does DOT restrict, log, and monitor access to the information security management systems (e.g., firewalls, vulnerability scanners, network sniffers, etc.)?

Accuracy

1. How often are training and data security and accuracy awareness provided to the employees?
2. How often does your organization consider annual review including third party providers upon which their information supply chain depends?
3. How often does your agency apply vulnerability management (the cyclical practice of vulnerability assessments) strategies in the system?
4. How often does your agency follow any data labeling standards (e.g. ISO 15489)?
5. How often does your agency apply vulnerability management (the cyclical practice of vulnerability assessments) strategies in the system?

Confidentiality

1. Whenever you want to install software or make any change in your computer User Access Control (UAC) pops up to confirm your action. How many computers are required to have UAC in your agency?
2. How often does your agency have control of data classification? (e.g. data asset identification, deploy classification, review, and declassification).
3. Does your agency require two-factor authentication for remote access? (e.g. token is used in addition to a username, and password).
4. Does your organization consider to have/ update the capability to use system geographic location as an authentication factor?
5. How often does your agency have controls in place ensuring timely removal of systems access that is no longer required?

In the next phase, we ask network security experts in DOT to answer these multiple-choice questions based on their experiment, knowledge and previous incidents of their network system. Depending on the type of question, the answers vary. For example, the answers for question 1 in Confidentiality factor can be as follow:

- No requirements
- Turned on for few computers
- Turned on for most computers
- Turned on for all computers, but users can click and pass it
- Turned on for all computers, but requires the administrator password to approve changes.

3.4 Multi-layered Fuzzy Logic

The main advantage of FL is its ability over other methodologies such as Decision Trees to model vagueness of concepts (e.g. low vulnerability or 'medium vulnerability'), because decision trees provide a discrete value based on input and output values. With decision trees, the actual function that makes a relation between vulnerabilities of the system with all available assets and potential risk is a non-linear function. But FL is not discrete and can be easier to analyze compared to other methodologies. Since the triangular model has the simplicity of MF compared to other models, we find it more appropriate and useful for vulnerability analysis. This model provides a smooth Fuzzy Inference System (FIS) to quantify a relationship between vulnerability attributes. This happens when input variables are converted to fuzzy variables in the fuzzification process. We use the most common properties of fuzzy logic for A_{fuzzy} and B_{fuzzy} as follow:

$$\mu_{A\cup B}(x) = min[\mu_A(x), \mu_B(x)] \,|\, x \in X$$
And
$$\mu_{A\cap B}(x) = min[\mu_A(x), \mu_B(x)] \,|\, x \in X$$

The defuzzification process converts the results of `if-then` rules to the crisp value. The most common defuzzification method is the centroid method [24]. This method considers the area of the fuzzy set, determines the center of the area and finally returns the corresponding crisp value of that [8, 10]. Figure 3 shows the overall perspective of the MFL method to measure the network system's vulnerability in DOT.

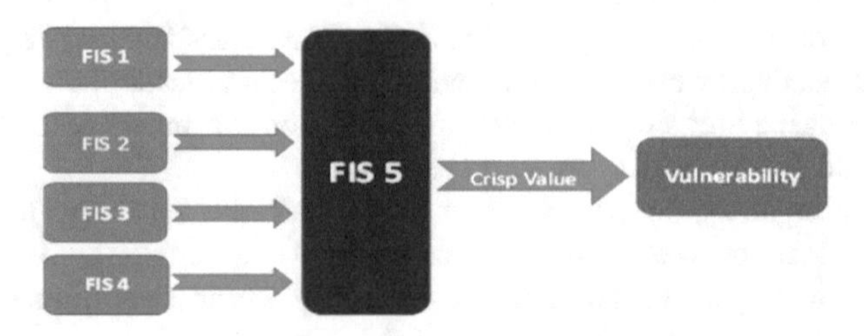

Fig. 3. Multi-layered FIS

Based on the answers received from the experts in the questionnaire, we categorize the results in the format of fuzzy subsets as 'very low', 'low', 'medium', 'high' and 'very high'. Afterward, based on the essence, characteristics, importance, and role of each security sub-factor in our security pattern, we divide them into two groups to formulate the final FIS using FL. Table 3 addresses this procedure and Table 4 expresses the description of conversion from expert answers to fuzzy subsets. For instance, after collecting data from the expert, their answer to each sub-factor is categorized to the appropriate Group 1 and Group 2 for vulnerability measurement. The answers may vary from 'very slow' to 'very high' (as shown in Table 3). Then, based on the weights assigned for each sub-factor in the questionnaire, they fall into each fuzzy subsets in each group. In this case, the factor Availability is the result of the combination of all sub-factors in Group 1 and Group 2.

Table 3. Fuzzy subset implementation for availability

Group 1	Fuzzy Subsets
Backup, D. Readiness, D.Recovery, W.Security.	Very low
	Low
	Medium
	High
	Very high

Group 2	Fuzzy Subsets
Redundancy, Monitoring, Q.Testing, L.Balancing.	Very low
	Low
	Medium
	High
	Very high

Factor	Fuzzy Subsets
Availability	Very low
	Low
	Medium
	High
	Very high

Table 4. Description of weighted questions to fuzzy subsets

Sub-factors	Description in fuzzy sets	
Backup	Very low	DOT never maintains a control framework to capture backup or in the best case, they have an annual backup program
	Low	DOT maintains a control framework to capture backup annually or twice a year
	Medium	DOT maintains a control framework to capture backup twice a year or at most quarterly
	High	DOT maintains a control framework to capture quarterly and at most monthly
	Very high	DOT maintains a control framework to capture backup monthly, biweekly, weekly or daily depends on the amount of data
Disaster readiness	Very low	DOT does not provide a disaster readiness plan to keep equipment away from a location subject to a high probability of natural disasters
	Low	DOT provides a disaster readiness plan to keep small pieces of equipment away from a location subject to a high probability of natural disasters
	Medium	DOT provides a disaster readiness plan to keep almost half of the equipment away from a location subject to a high probability of natural disasters
	High	DOT provides a disaster readiness plan to keep all types of equipment away from a location subject to a high probability of natural disasters

(*continued*)

Table 4. (*continued*)

Sub-factors	Description in fuzzy sets	
	Very high	DOT provides a disaster readiness plan to keep all types of equipment away from a location subject to a high probability of natural disasters with redundant supplementary equipment
Disaster recovery	Very low	DOT does not have any plan or has a weak plan to have a third party to minimize the impact of natural disasters or unauthorized access
	Low	A third party rarely has access to DOT's information system to minimize the impact of natural disasters with compensating control, unauthorized access in emergency cases
	Medium	A third party sometimes has access to DOT's information system and following a coordinated application of resources to minimize the impact of natural disasters with compensating controls, unauthorized access in emergency cases
	High	A third party regularly has access to DOT's information system and following a coordinated application of resources to minimize the impact of natural disasters with compensating controls, unauthorized access in emergency cases
	Very high	A third party always has access to DOT's information system and following a coordinated application of resources to minimize the impact of natural disasters with compensating controls, unauthorized access in emergency cases
Wireless Security	Very low	DOT does not follow an established policy to protect wireless network environments
	Low	DOT does not follow an established policy to protect wireless network environments, but sometimes takes it into account for consideration
	Medium	DOT sometimes follows an established policy to protect wireless network environments such as perimeter firewalls and strong encryption for authentication
	High	DOT follows an established policy to protect wireless network environments such as perimeter firewalls and strong encryption for authentication (but is not strict on the policy)
	Very high	DOT always follows an established policy to protect wireless network environments such as perimeter firewalls and strong encryption for authentication
Redundancy	Very low	DOT does not have any protection plan to take care of all equipment from utility service outages
	Low	DOT a weak protection plan to take care of some equipment from utility service outages
	Medium	DOT puts protection measures to take care of almost half of the equipment from utility service outages
	High	DOT puts protection measures to take care of the majority of the equipment from utility service outages
	Very high	DOT puts protection measures to take care of all the equipment from utility service outages
Monitoring	Very low	DOT does not have any industry-standard monitoring and alerting on devices
	Low	DOT provides a weak plan to perform industry-standard monitoring on devices
	Medium	DOT regularly perform industry-standard monitoring and alerting on devices that process and store sensitive information
	High	DOT usually perform industry-standard monitoring and alerting on devices that process and store sensitive information but is not strict on that
	Very high	DOT always perform industry-standard monitoring and alerting on devices that process and store sensitive information

(*continued*)

Table 4. (*continued*)

Sub-factors	Description in fuzzy sets	
Quality testing	Very low	DOT does not have any plan to provide policies and mechanisms of quality testing for software
	Low	DOT provides a poor plan to follows policies and mechanisms of quality testing for software
	Medium	DOT regularly follows policies and mechanisms of quality testing including unit, integration, system and acceptance testing for released software versions
	High	DOT usually provides and follows all policies and mechanisms of quality testing including unit, integration, system and acceptance testing for released software versions
	Very high	DOT always provides and follows all policies and mechanisms of quality testing including unit, integration, system and acceptance testing for released software versions
Load balancing	Very low	DOT does not have any load balancing plan for network security
	Low	DOT rarely provides a load balancing plan for network security
	Medium	DOT sometimes provides and follows a plan of load balancing for network security
	High	DOT usually provides and follows a plan of load balancing for network security
	Very High	DOT always provides and follows a strong plan of load balancing for network security such as distributing network traffic across multiple servers

4 Implementation

Since we cannot address the implementation of all security factors in one paper, here we address only the Availability security factor as part of the MFL approach. Figure 4 shows the procedure of FIS for Availability. We define the measurement procedure in a way that score 10 indicates the highest Availability and 0 indicates the lowest one in DOT's network system.

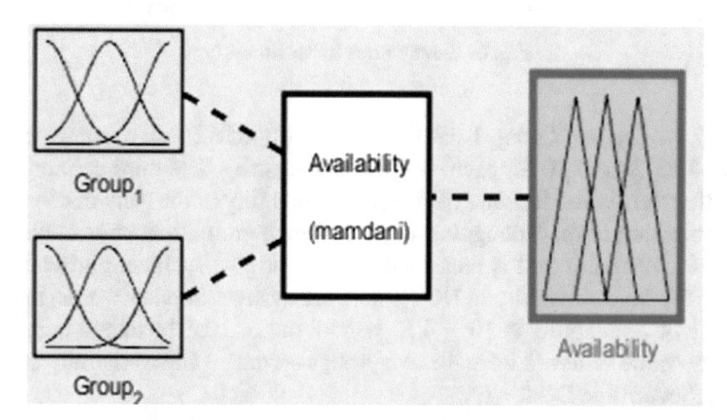

Fig. 4. FIS for availability

Based on the model we represented above, we apply `if-then` rules consisting of the antecedent 'if' and the consequent 'then' to integrate all sub-factors of each category that are linguistic-based components of vulnerability measurement. Fuzzy subsets and MF for Group 1 is shown in Fig. 5. In addition, Fig. 6 shows 10 out of 25 rules associated with Availability in Mamdani [11] model. Each rule can be defined in the interval of [0, 1] variation; however, in our research, all rules have the same weight of 1. However, the actual weights of each security sub-factor are derived from experts and analyses in the questionnaire.

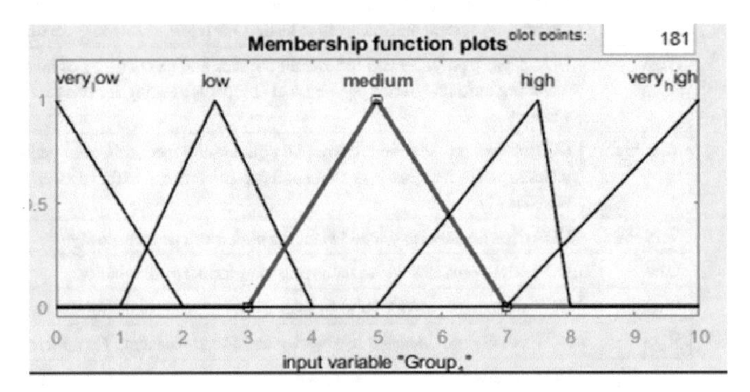

Fig. 5. Fuzzy subsets and MF

1. If (Group_1 is very_low) and (Group_2 is very_low) then (Availability is very_low) (1)
2. If (Group_1 is very_low) and (Group_2 is low) then (Availability is very_low) (1)
3. If (Group_1 is very_low) and (Group_2 is medium) then (Availability is low) (1)
4. If (Group_1 is very_low) and (Group_2 is high) then (Availability is low) (1)
5. If (Group_1 is very_low) and (Group_2 is very_high) then (Availability is medium) (1)
6. If (Group_1 is low) and (Group_2 is very_low) then (Availability is very_low) (1)
7. If (Group_1 is low) and (Group_2 is low) then (Availability is low) (1)
8. If (Group_1 is low) and (Group_2 is medium) then (Availability is low) (1)
9. If (Group_1 is low) and (Group_2 is high) then (Availability is medium) (1)
10. If (Group_1 is low) and (Group_2 is very_high) then (Availability is medium) (1)

Fig. 6. Fuzzy rules for availability

In Fig. 7 we see that Group 1 has the Availability of 6.19 on average and Group 2 shows it as 3.81 (out of 10 for each) for network security. The third column represents the result of each rule and it shows the overall Availability of the Network System based on the combination of the Triangular method of each group, that after implementation of all rules for Availability it is measured as 4.52 out of 10 which represents the level of vulnerability for Availability in DOT's network system. In other words, the potential vulnerability of Availability is $10 - 4.52 = 5.48$ out of 10. The output curve in Fig. 7 addresses the value achieved from the two groups comprise all Availability components of Network Security in DOT.

As shown in Fig. 8, in the vertical axis (Availability), is represented in the range of 0 to 9 in which 0 indicates the least Availability (maximum vulnerability) and 9 indicates the maximum Availability (least vulnerability). The maximum value for Vulnerability is not predetermined, it is handled by FIS through a calculation based on MF.

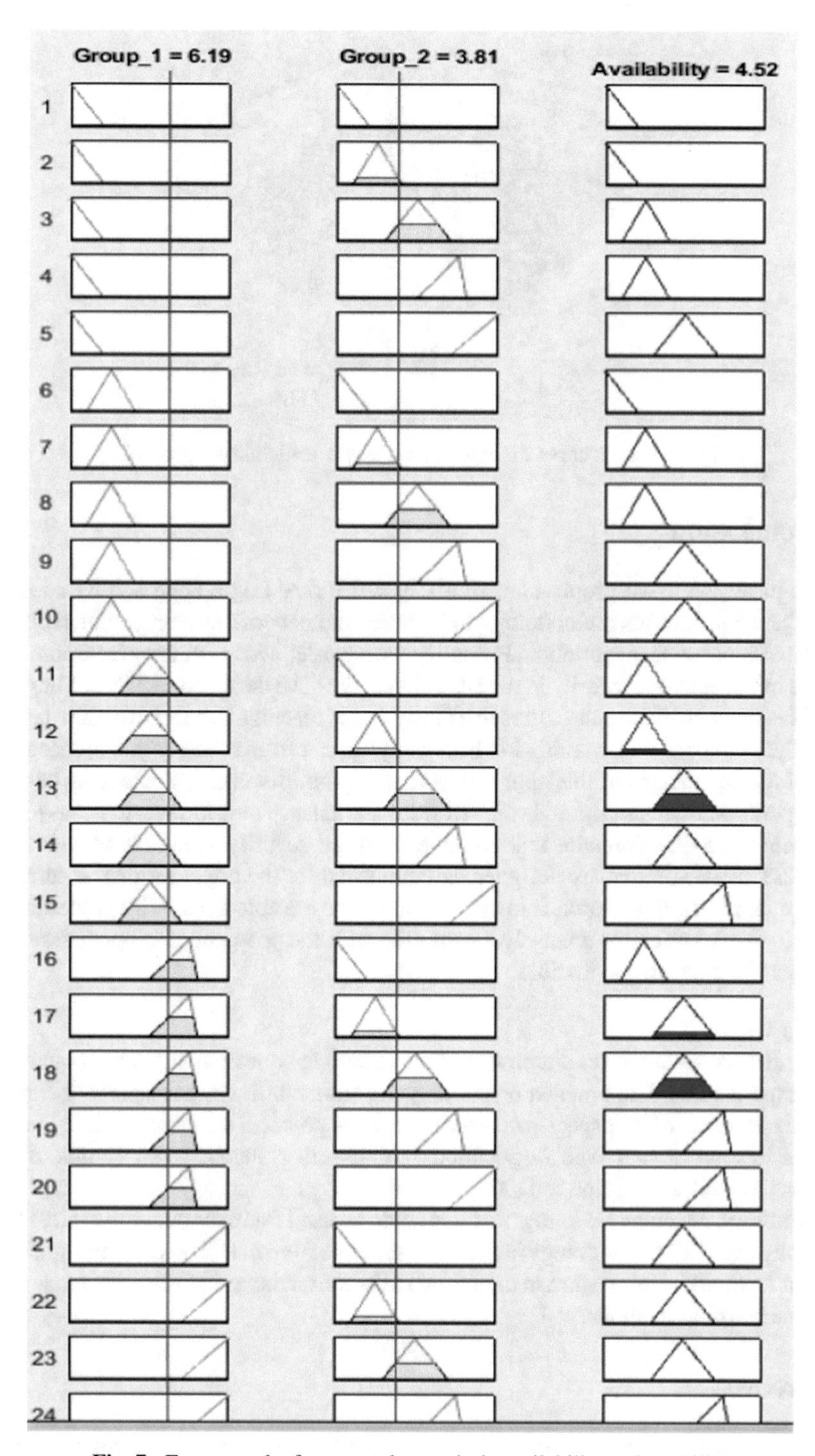

Fig. 7. Fuzzy results for network security's availability-vulnerability

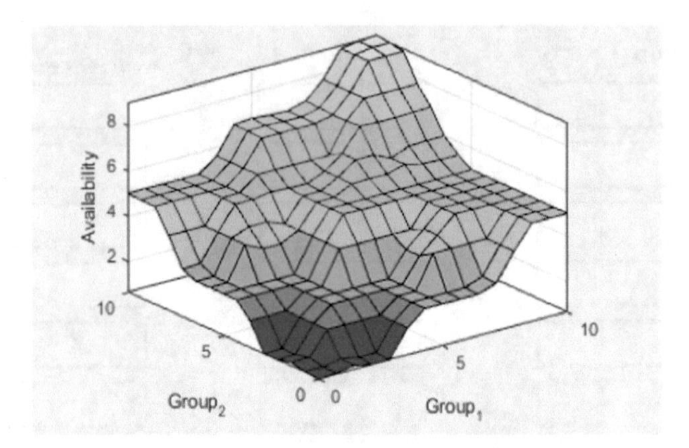

Fig. 8. Output curve for network security's availability-vulnerability

5 Conclusion

In this paper, we have proposed a Multi-layered Fuzzy Logic approach based on the Goal Question Metrics methodology to quantify the network system's vulnerability in the Department of Transportation. Primarily, we have defined all security factors and sub-factors of the network security in the DOT. Then, we have designed a subset of linguistic variables for each of the sub-factors in a Fuzzy Logic machine in order to obtain the exact numerical values of vulnerability in a backward manner from successors to predecessors. One of the advantages of this approach is that it quantifies each vulnerability based on the expert's knowledge through real-world experience. And, in fact, it is based on a comprehensive questionnaire that complies with all security standards of ISO 27001 and NIST SP800-53 that are designed and published for the organizations' security. By virtue of applying this approach in cybersecurity, we are able to quantify vulnerabilities in the network system as precisely as possible with the prospect of keeping assets and information as secure as possible.

Future Work

In the next phase of the research, we will distribute the questionnaire to computer network experts in DOT to broadening not only the vulnerability measurement of Network Security but also other aspects of security such as physical security, Mobile Devices, Human Resources, and Web Applications. Long term plans are to apply this to other domains as well, in addition to DOT.

Moreover, Machine Learning methods could be used for individual sub-factors in the hierarchy tree. Our methodology is currently separate from machine learning algorithms, but that it would be interesting in the future to do comparisons between this method and machine learning methods.

References

1. Hayden, L.: IT Security Metrics: A Practical Framework for Measuring Security & Protecting Data. McGraw Hill, New York (2010)

2. Karabacak, B., Sogukpinar, I.: ISRAM: information security risk analysis method. Comput. Secur. **24**(2), 147–159 (2005)
3. Basili, V.R., Green, S.: Software process evolution at the SEL. Foundations of Empirical Software Engineering, pp. 142–154
4. Zimmermann, H.-J.: Fuzzy Sets, Decision Making, and Expert Systems. Springer, Dordrecht (1987)
5. Zadeh, L.: Fuzzy sets. Inf. Control **8**(3), 338–353 (1965)
6. Bibliography on fuzzy sets and their applications. In: Fuzzy Sets and Their Applications to Cognitive and Decision Processes, pp. 477–496 (1975)
7. Pedrycz, W.: Why triangular membership functions? Fuzzy Sets Syst. **64**(1), 21–30 (1994)
8. Kacprzyk, J.: Group decision making with a fuzzy linguistic majority. Fuzzy Sets Syst. **18**(2), 105–118 (1986)
9. Karnik, N., Mendel, J., Liang, Q.: Type-2 fuzzy logic systems. IEEE Trans. Fuzzy Syst. **7**(6), 643–658 (1999)
10. Leekwijck, W.V., Kerre, E.E.: Defuzzification: criteria and classification. Fuzzy Sets Syst. **108**(2), 159–178 (1999)
11. Mamdani, E.: Advances in the linguistic synthesis of fuzzy controllers. Int. J. Man Mach. Stud. **8**(6), 669–678 (1976)
12. Erturk, E., Sezer, E.A.: Software fault prediction using Mamdani type fuzzy inference system. Int. J. Data Anal. Tech. Strat. **8**(1), 14 (2016)
13. Sugeno, M.: An introductory survey of fuzzy control. Inf. Sci. **36**(1–2), 59–83 (1985)
14. Pfleeger, C.P., Pfleeger, S.L., Margulies, J.: Security in computing. Pearson India Education Services, India (2018)
15. Anton, P.S., Anderson, R.H., Mesic, R.: Finding and Fixing Vulnerabilities in Information Systems: The Vulnerability Assessment and Mitigation Methodology. RAND Corporation, Santa Monica (2004)
16. Measuring Operational Risk Using Fuzzy Logic Modeling. Measuring Operational Risk Using Fuzzy Logic Modeling | Expert Commentary | IRMI.com. https://www.irmi.com/articles/expert-commentary/measuring-operational-risk-using-fuzzy-logic-modeling. Accessed 24 Jan 2020
17. Zhao, D.-M., Wang, J.-H., Ma, J.-F.: Fuzzy risk assessment of the network security. In: 2006 International Conference on Machine Learning and Cybernetics (2006)
18. Lee, M.-C.: Information security risk analysis methods and research trends: AHP and fuzzy comprehensive method. Int. J. Comput. Sci. Inf. Technol. **6**(1), 29–45 (2014)
19. Watkins, L., Hurley, J.S.: Cyber maturity as measured by scientific-based risk metrics. J. Inf. Warf., 60–69
20. The ISO 27001 Risk Assessment: Information Security Risk Management for ISO 27001/ISO 27002, 3rd edn., pp. 87–93 (2019)
21. J. T. Force: Security and Privacy Controls for Information Systems and Organizations, CSRC, 15 August 2017. https://csrc.nist.gov/publications/detail/sp/800-53/rev-5/draft
22. Shepperd, M.: Practical software metrics for project management and process improvement. Inf. Softw. Technol. **35**(11–12), 701 (1993)
23. Shojaeshafiei, M., Etzkorn, L., Anderson, M.: Cybersecurity framework requirements to quantify vulnerabilities based on GQM. Springer, 04 June 2019. https://link.springer.com/chapter/10.1007/978-3-030-31239-8_20
24. Ngai, E., Wat, F.: Fuzzy decision support system for risk analysis in e-commerce development. Decis. Support Syst. **40**(2), 235–255 (2005)

Cyber Security Technology

Challenges of Securing and Defending Unmanned Aerial Vehicles

William Goble[1], Elizabeth Braddy[1], Mike Burmester[1(✉)], Daniel Schwartz[1], Ryan Sloan[1], Demetra Drizis[1], Nitish Ahir[1], Melissa Ma[1], Matt Bays[2], and Matt Chastain[2]

[1] Department of Computer Science, Florida State University, Tallahassee, FL 32304, USA
{wg13b,cjb17e,ras18e,dmd16e,ndal6b,mm16bn}@my.fsu.edu, {burmester,schwartz}@cs.fsu.edu

[2] Naval Surface Warfare Center, Panama City Division, Panama City, FL 32407, USA
{matthew.bays,matthew.chastain}@navy.mil

Abstract. There are increasing concerns that foreign manufactured unmanned aerial systems may leak sensitive data to their manufacturers, particularly since such systems are used for reconnaissance and surveillance of critical infrastructure, for monitoring/managing industrial incidents, for tracking terrorist attacks, and more generally in applications that involve homeland/national security. In this paper we investigate the challenges of securing and defending such systems, focusing on civilian Group 1 (small) drones (quadcopters). We propose a solution based on an architecture that complies with the policies and standards of the Committee on National Security Systems for the Cybersecurity of Unmanned National Systems CNSSP 28, in which software components are adapted/modified appropriately, and security policies/mechanisms are enforced. Protection builds on isolation, encapsulation, and the use of cryptographic tools, with performance constraints expressed in terms of computation (power) and latency.

Keywords: Unmanned Aerial Systems · Drones · CNSSP 28 Policies · Cybersecurity · Stream authentication · XSS vulnerabilities

1 Introduction

Advances in sensor and energy technologies, as well as communication technologies, have made unmanned aerial systems (UAS) ideally suited for autonomous in-depth surveillance and reconnaissance, enabling a vast range of remote sensing services and applications. Many of these services involve sensitive data related

---

M. Burmester–Any opinions, findings, and conclusions or recommendations expressed in this material are those of the authors and do not necessarily reflect the views of the Naval Engineering Education Consortium and the Naval Surface Warfare Center.

K.-K. R. Choo et al. (Eds.): NCS 2020, AISC 1271, pp. 119–138, 2021.
https://doi.org/10.1007/978-3-030-58703-1_8

to critical infrastructure (power grids, transportation, agriculture, critical manufacturing, customs and border protection, etc.), industrial incidents and tracking terrorists attacks. This raises security concerns and the possibility that sensitive data may leak. The U.S. Department of Homeland Security recently warned U.S. firms of the risks to company data from Chinese-made drones [27], and in particular that there were strong concerns that drone technologies are used to leak sensitive data (intelligence). This has led to the Interior Department grounding its entire fleet [30], and the Marines Corps shelving small drones [10].

In this paper we investigate the challenges of securing commercial UASs. We focus on Group 1 unmanned aerial vehicles (UAV, typically: small battery powered quadcopters, 30 min flight time, 5 mi operating range, 1 kg weight), and propose a solution for which, by appropriately adapting/modifying software components and enforcing security policies/mechanisms, we can design a secure operations architecture that complies with the CNSSP 28 Policies for Cybersecurity of unmanned national systems of the Committee for National Security Systems [4]. Protection is based on isolation, encapsulation, and the use of cryptographic tools.

Main Challenges: (*i*) Understand the threat model of UASs and identify the vulnerabilities of their implementations, in particular the cyber/physical threats to heterogeneous wireless networks, and (*ii*) develop methodologies and cyber forensics that identify UAS malware (e.g., XSS exploits).

Main Contributions: Develop methodologies, policies and mechanisms that protect and defend UASs, and in particular, a CNSSP 28 [4] compliant framework that protects and defends (*i*) the hardware and software controls of UASs, (*ii*) their communication channels, (*iii*) transmitted data that is not releasable, and that offers real-time ongoing security awareness. Protection uses virtualization technologies (for domain separation, process isolation and resource encapsulation) and cryptographic tools (for integrity, authentication and privacy), and addresses transmission security (privacy exploits other than cryptanalysis), emissions security (privacy/integrity exploits based on electromagnetic emanations) and physical security.

The rest of this paper is structured as follows. In Sect. 2 we overview recent literature on the challenges of securing UASs. In Sect. 3 we discuss the policies and technologies we shall be using: the CNSSP 28 policies for cybersecurity of unmanned national systems, the threat model and the cost of protecting UASs, the MAVLink communication protocols, and virtualization technologies. Section 4 is devoted to our approach that is based on a CNSSP 28 compliant architecture. To address transmission, emissions and physical security requirements, we describe a secret-key variant of Wyner's wiretap channel, that can be used for offensive and defensive security. It can also be used to bypass crypto during operations. Section 5 discusses future work and concludes.

2 Background

Although there is widespread use of drones for commercial applications, little attention has been given to their security. Of the publications that address security, many are essentially proposals, or proofs of concept. MAVLink 1.0 [15] is the first protocol designed for UAV applications, but this focusses on efficiency and reliability. A later version, MAVLink 2.0 [16] addresses integrity/authentication. There is also a draft, sMAVLink [28], that considers privacy. Davanian et al. [5] proposed a MAVLink implementation that offers some security whilst imposing no performance overhead by using a "moving target defense" and "software diversity" approach, in which system aspects keep changing to vary the attack surface and make it harder for an attacker to exploit vulnerabilities.

Several publications discuss signal spoofing and hacking attacks on drones in which the attacker gets control of flight operations or manipulates the drones' autopilot system. Arthur [1] proposes using adaptive intrusion detection methodologies to identify intruders. Krishna et al. [12] analyze GPS jamming and spoofing attacks as well as de-authentication attacks in which the attacker targets the control stream between the drone and ground control. Ward [34] discusses the legal, technical and practical challenges of countering commercial drone threats to national security, and points out that although there is no perfect interdiction solution for mitigating rogue UASs, hacking and/or spoofing may be the most promising solution (with the least potential for collateral damage). Several publications propose using Verification and Validation (V&V) procedures to assure UAV compliance with safety standards. Torens et al.[31] consider a V&V approach for a sampling-based automated mission planner for UAVs. Dill et al. [6] describe a prototype assured compliance system for small to midsize UASs, called Safeguard, that constrains the drone's operation to mitigate safety risks associated with avionics systems failures.

Air traffic management is of critical importance to maintain safe and collision-free operations. Guvenc et al. [11] address the issue of detecting, localizing and tracking unauthorized drones and jammers, and present some experimental and simulation results for radar-based range estimation and receding horizon tracking. Srinivas et al. [29] consider the problem of authorized users (external parties) being able to access real-time data from drones in an Internet of Drones environment and propose an (anonymous) authentication protocol. Finally, Won and Bertino [36] discuss ways of protecting data collected by drones and secret keys, by integrating techniques for software-based attestation, data encryption and secret key protection. In their approach free memory space is filled up with pseudo-random numbers that are then used to encrypt data as a stream cipher. This assures that an attacker that can see memory space cannot obtain information about collected data.

3 CNSSP 28 Policies, Threat Model, MAVLink and Virtualization

Our goal is to design a lightweight unmanned aerial systems (UASs) architecture for a typical civilian drone that provides real-time protection with ongoing security awareness, by adapting/modifying its software components. In this section we discuss the CNSSP 28 security policies and standards for UAS, the threat model, the communication protocols and the virtualization tools that we shall use.

3.1 CNSSP 28: Cybersecurity of Unmanned National Systems

CNSSP 28 [4] is a policy document issued by the Committee on National Security Systems for Cybersecurity of unmanned national systems. Such systems operate in all physical environments (underwater, surface and space), are controlled remotely and autonomously, and support diverse and complex missions. Below we list some of the policies we shall use to secure our particular applications.

A. Unmanned national security systems must be designed to operate through cybersecurity-related attacks (wireless or wired) to the extent necessary to execute assigned missions.
B. NSA-approved cryptographic algorithms, techniques and associated security architectures must be used whenever needed. At a minimum they must be used to:
   a Authenticate and encrypt all command control (C2) information transmitted over communication links and include link loss countermeasures to provide predictive operations and enable C2 link recovery;
   b Encrypt all data not releasable to unauthorized personnel that is transmitted over communication links;
   c Authenticate & encrypt all communication links used with remotely-controlled mission termination system (self-destruct command);
   d Implement countermeasures to prevent or delay exploitation and provide tamper evidence of devices and systems containing data not releasable to unauthorized personnel;
   e Unmanned national security systems that use commercial solutions for receiving, disseminating, processing and storing not releasable information must follow NSA approved zeroization procedures;
   f Data-at-rest encryption to ensure hard drives and other non-volatile storage devices should be not exploitable if captured;
C. Develop communication security (COMSEC) measures and procedures to deny unauthorized users' information and ensure the authenticity of such communication, including cryptographic, transmission, emissions and physical security.
D. Design crypto-bypass to regain command control of unmanned NSS when encrypted communications are lost, or in emergencies, when circumstances dictate such actions. The design must minimize the possibility of bypass activation due to malicious activity or failure.

E. Develop plans for recovery and protection of any classified material and equipment, components, or keying material of the unmanned system that loses positive C2 and terminates at an unsecured site/location.

3.2 Threat Model and Cost of Protecting UAVs

The goal of the attacker is to exploit system vulnerabilities, while the goal of the system designer is to eliminate vulnerabilities. Despite the apparent symmetry, the threat model is highly asymmetric: the attacker just needs to find one exploit, whereas the designer must eliminate exploits before the attacker finds them, or mitigate their impact (for system resilience). For security analysis the adversary is modeled by a probabilistic polynomial time Turing Machine that controls the communication channels, and may eavesdrop, block, modify and/or inject messages in any communication between parties (Dolev-Yao model [8]). The adversary may also corrupt stored data. Security is typically defined in terms of indistinguishably (semantic security). For complex systems such as UASs, security should be holistic, and an ongoing process that involves continuous monitoring, risk awareness/assessment and tolerance.

Experiments have shown that up to 95% of the energy of a UAV is consumed to keep it airborne and support its movement [26]. Consequently for battery powered UAVs, the time complexity of protection mechanisms is the predominant factor for assessing their effectiveness. We note however that traditional UAVs rely on line-of-site (LoS) communication over unlicensed spectrum which has low data rate, is unreliable, insecure and prone to interference. Communication complexity can therefore be a major challenge. Similarly for lightweight UAV applications (e.g., with a 1GHz ARM Cortex-A8 processor) the computational complexity can be challenging. We shall therefore be using protocols (such as MAVLink) that address such concerns.

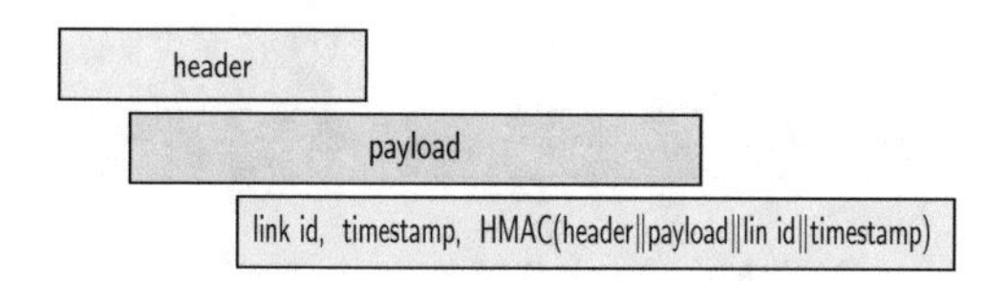

Fig. 1. A MAVLink 2.0 message packet: the header, payload, and timestamped HMAC.

3.3 The Micro Air Vehicle Communication Protocols

MAVLink is an open source cross-platform lightweight communication protocol for UAS applications that serializes messages into a specific binary form. There are two versions: MAVLink 1.0 [15], MAVLink 2.0 [16], and a draft version sMAVLink [28]. Combined with the mavros Robotics OS (ROS) [17], MAVLink enables communication between the autopilot module and the ground control station.

MAVLink 1.0 (2009) is designed to optimize efficiency and reliability, in non-adversarial settings where faults are random, or transient, resulting from weak signals or sensor failure (it uses cyclic redundancy checks (CRC) to detect errors during transition). A later version, MAVLink 2.0 (2017) uses timestamped hash-based message authentication codes (HMAC) for integrity and authentication. The draft version sMAVLink is a security-enhanced version that uses symmetric key authenticated encryption with associated data for confidentiality and integrity.

MAVLink 2.0 Packets (Fig. 1). Message packets have a 10-byte *header*, an up to 255-byte *payload*, and a 13-byte timestamped HMAC.[1] The header has 8 fields: a start-of-frame STX; a payload-length LEN; an incompatibility flag INC; a computability flag CMP; a packet-sequence SEQ, a SYS ID (sender), a COMPonent ID, and a MSG ID that indicates the type of message: each is 1 byte, except the last that is 3 bytes. We distinguish *state* messages and *command* messages. State messages contain information about the system and include: *heartbeat*, *system state* messages, and *global position* messages. Heartbeat messages (9 bytes) are used for *liveness* checks and are sent at regular intervals (1 Hz). Heartbeat packets have 6 fields containing information about the unmanned system (Fig. 2): type of vehicle (1 byte); type of autopilot (1 byte); base mode (1 byte); custom mode (4 bytes, autopilot specific flags); system status (1 byte); and mavlink version (1 byte) [14]. System state messages contain data about onboard control sensors: which sensors are enabled/disabled, battery levels, etc. These messages also contain information about communication errors and dropped frames. Global position messages contain information about the latitude, longitude and altitude of the vehicle. Command messages contain, command and control (C2) data and mission termination data.

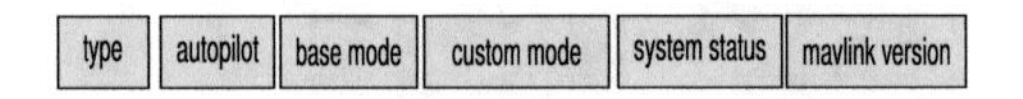

Fig. 2. A MAVLink 9-byte heartbeat packet.

The timestamped HMAC of MAVLink 2.0 has three fields (Fig. 1): a 1-byte link id that identifies the type of link used to send the packet; a 6-byte timestamp (using a 10 ms resolution); and the first 6 bytes of the hash-based message authentication code HMAC-SHA256 on the concatenation of the header, payload, link id and timestamp. The 256-bit secret key is shared by the drone and the GCS. The time taken to compute a MAVLink HMAC is approximately 26 µs per packet when using a Pixhawk 1 flight controller (CPU: 180 MHz ARM, Cortex M4; SRAM: 256 KB [32]).

[1] The CRC for error detection of MAVLink 1.0 is not needed when HMACs are used.

3.4 Virtualization Technologies for Isolation, Separation and Encapsulation

Virtualization is an application program interface (API) that partitions the execution environment of a computer system into multiple isolated virtual environments [25]. With *hypervisor* virtualization, partitioning takes place at the hardware level (Fig. 3a), while with *container* virtualization it takes place at the host OS level (Fig. 3b). Hypervisor virtualization uses an encapsulating software layer, the *hypervisor* or *virtual machine monitor* (VMM), to partition the (physical) host computer into isolated logical virtual machines (VM) that run in environments created to address specific tasks. The hypervisor creates and runs the VMi and their operating systems OSi ($i = 1, \ldots, n$) on the host computer, and surrounds (underlies) the OSi, providing all required inputs, outputs and hardware resources of the actual computer (CPU, memory, I/O devices, etc.), to maintain domain separation, process isolation (even though the VMs share resources of the host) and resource encapsulation (by bundling in a software package the virtual resources, the VMi, the OSi and their applications). Hypervisor virtualization decouples system hardware from software running on it and system resources, with all instructions carried out in a "sandboxed" environment. Depending on the implementation, we get both flexibility (multiple OSi can run simultaneously on the same machine), and security (by restricting the impact of malicious behavior). Container virtualization (Fig. 3b) is a lightweight abstraction that uses the host kernel to run multiple isolated virtual environments, called *containers*.

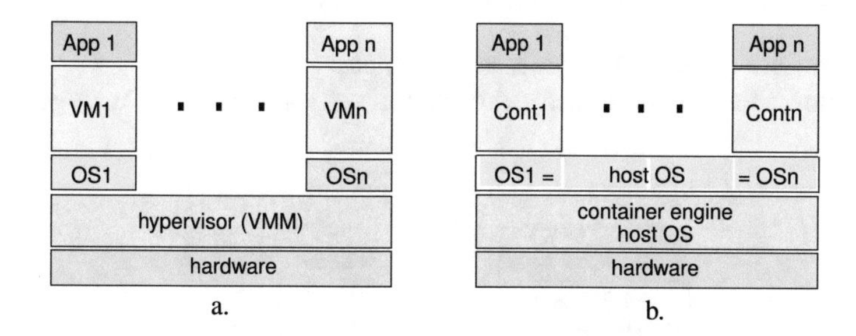

Fig. 3. Architectures of (a) hypervisor-based and (b) container-based virtualizations.

Hypervisor virtualization techniques are claimed to be more secure than container techniques as an extra layer of isolation is added between the host and application. For example, a malicious App 1 running on VM1 (Fig. 3a) can only interact with the kernel of VM1, not the host kernel (e.g., cannot infect OSn). However container virtualizations (Fig. 3b) share the host OS kernel that makes it possible for malicious behavior to spread.

There are several ways to harden container encapsulation. The security enhanced SELinux [35] and AppArmor module for Docker containers provides

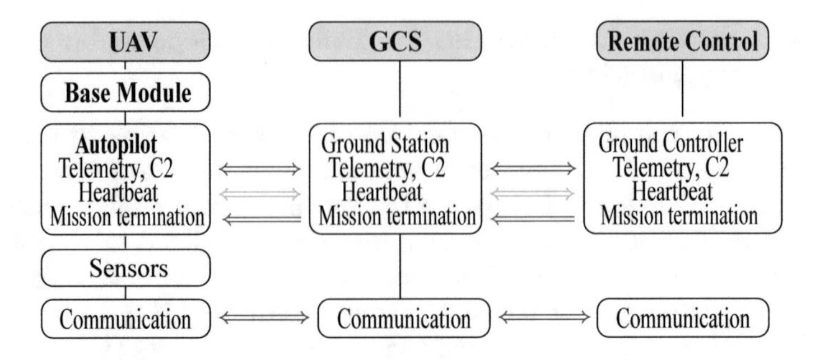

Fig. 4. The components of an UAS and their links.

policy mechanisms for mandatory access control (MAC) and discretionary access control (DAC) [7]. With these security modules we get a more flexible and robust encapsulation that provides better performance than hypervisor virtualization, with higher density virtual container environments [38]. For our application we shall use the Docker Engine container API [35] with the Linux security module [7] (and AppArmor).

4 A CNSSP 28 Compliant Framework for UASs

An unmanned aerial system (UAS) consists of: an unmanned aerial vehicle (UAV, drone), a ground control station (GCS) and, one or more Remote Controls (control interfaces). The basic components of the UAV are the,

- Base Module (BM): microprocessor, OS, etc.
- Autopilot Module (Avionics): ArduPilot software suite for navigation control [2].

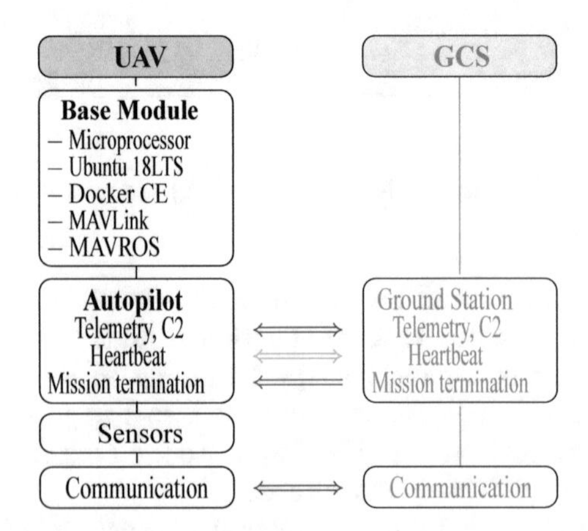

Fig. 5. A CNSSP 28 compliant framework for the UAV.

- Sensors Module: accelerometer, gyroscope, compass, GPS, etc., and
- Communication links: telemetry, C2, heartbeat, mission termination, etc.

Figure 4 illustrates the structure of an UAS. The BM contains the flight controller of the UAS: a microprocessor and the OS (firmware, middleware and software) that controls the avionics and communication links (telemetry, heartbeat, mission termination and C2 channels). The Autopilot Module converts state (navigational) data (telemetry streams: position, velocity, altitude, etc.) and command and control data (C2 data) it receives from the GCS to commands for engines, flaps, rudder, stabilizer and spoilers, and sends state data to the GCS.

The Sensors Module consists of the sensory equipment with integrated functionalities (accelerometer, gyroscope, compass, GPS, etc.). Navigation is assured by an inertial navigation system whose drift is compensated using global position system (GPS) measurements. Liveness is checked using heartbeat signals. The BM is securely paired with the GCS, which in turn is securely paired with the Remote Control (secure device pairing is a process of bootstrapping a secure channel between two devices [13]).

4.1 A CNSSP 28 Framework for UAV

The framework for an UAV is shown in Fig. 5. It consists of:

- An appropriate microprocessor [17];
- A Docker CE (Community Edition [35]) API: daemon, objects (containers, images, services), registries and tools [7];
- The Linux kernel security module AppArmor, and the secure computing module seccomp (for MAC and DAC policy control) [7];
- The Ubuntu 18LTS Linux operating system [33];
- The MAVLink [15] communication protocol between the Autopilot Module and GCS;
- The MAVROS [17] Robotic OS on MAVLink.

In the following sections we discus the mechanisms and policies needed to protect the UAV and the MAVLink messages for CNSSP 28 compliance.

4.2 Securing UAVs: Isolation and Encapsulation (CNSSP 28 Policies: A, C)

To address the threat of the drone's host hardware and firmware being compromised, we shall use virtualization technologies.

Domain Separation, Process Isolation and Resource Encapsulation. The hardware (microprocessor) and firmware of the drone control the BM as well as the avionics, telemetry, heartbeat, C2, sensor and mission termination channels. To protect these applications the distinct functionalities of the drone are processed in isolated Docker containers. We propose two approaches: the first uses one container while the second uses three containers.

One Container: this controls the BM, the Autopilot (avionics, telemetry, etc.) and all communication channels.

Three Containers: a container that controls the BM and Autopilot, a container that controls all communication channels except mission termination, and a container that controls the mission termination process.

The security objective of virtualization is to isolate compromised hardware/software and contain malware infection outbreaks and attacks. The one container solution isolates drone applications from potentially compromised underlying hardware and software. The three container solution extends protection to applications in which containers get compromised. For example, mission termination instructions are only used as a last resort, typically when the behavior of a drone indicates that it is compromised. So the container that manages termination instructions is unlikely to get compromised. However other containers may get compromised. If a mission termination instruction is sent (e.g., because of unexpected drone behavior), it must be executed (even if compromised containers try to prevent this). Consequently, such instructions should be protected and have exclusive control of the components needed for their execution (MAC policy mechanisms of the Linux security module are used to enforce access).

4.3 Securing MAVLink Messages

There are two types of MAVLink messages that need to be protected: state messages that contain information about the state of the system (e.g., heartbeat messages) and mission messages (e.g., C2 messages).

Securing Heartbeat Messages (CNSSP 28: A, Ba, Bd, C, D). The MAVLink heartbeat (Fig. 2) is not authenticated by the drone and can easily be forged. For reliable liveness detection, heartbeat messages can be authenticated using 13 byte MAVLink timestamped HMACs. For lightweight applications the time complexity of an HMAC is roughly 26 µs (Sect. 3.3). However the added communication cost is more than double that of the heartbeat (9 bytes). This may cause delays, and adversely affect the operation of the drone. Another approach is to use stream authentication.

Stream Authenticated Heartbeat. In this protocol the drone and GCS share a pseudo-random number generator (PRNG) and exchange successive numbers $n_0, \ldots, n_{2j}, n_{2j+1}, \ldots$, drawn from it. The numbers sent authenticate the parties. We explain the protocol in more detail.

Each party (GCS and drone) has a copy of the same PRNG, $g(state)$ (hereafter abbreviated as g), which are (loosely) synchronized, with numbers drawn from g as needed for successive authentication. For efficiency and resiliency, at the j-th exchange, the GCS and drone have a list of numbers drawn from g: $list_{gcs} = [n_{2j}, n_{2j+1}, n_{2j+2}]$, $list_d = [n_{2j-2}, n_{2j-1}, n_{2j}, n_{2j+1}]$. Initially ($j = 0$) the shared PRNGs are synchronized, n_0, n_1, n_2 are drawn, $list_{gcs}$ is set to $[n_0, n_1, n_2]$, and $list_d$ is set to $[\lambda, \lambda, n_0, n_1]$, where λ is the empty string. When

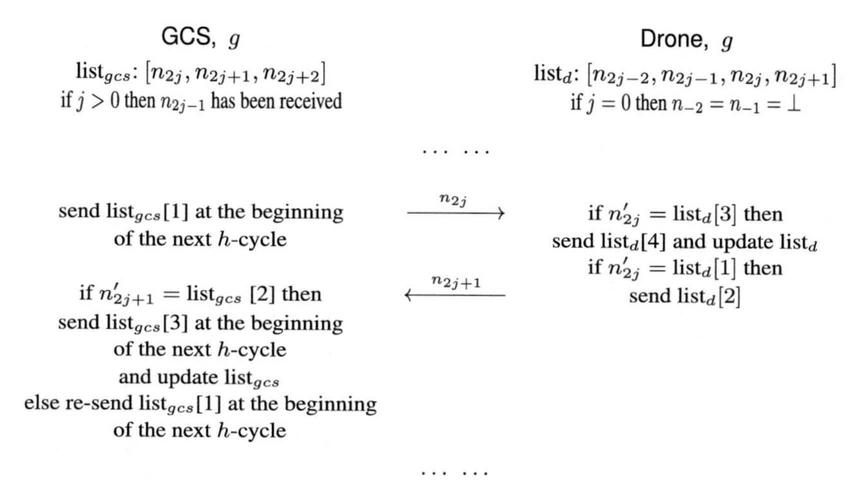
GCS, g
list_{gcs}: $[n_{2j}, n_{2j+1}, n_{2j+2}]$
if $j > 0$ then n_{2j-1} has been received
Drone, g
list_d: $[n_{2j-2}, n_{2j-1}, n_{2j}, n_{2j+1}]$
if $j = 0$ then $n_{-2} = n_{-1} = \perp$
send $\text{list}_{gcs}[1]$ at the beginning of the next h-cycle
n_{2j}
if $n'_{2j} = \text{list}_d[3]$ then send $\text{list}_d[4]$ and update list_d
if $n'_{2j} = \text{list}_d[1]$ then send $\text{list}_d[2]$
n_{2j+1}
if $n'_{2j+1} = \text{list}_{gcs}$ [2] then send $\text{list}_{gcs}[3]$ at the beginning of the next h-cycle and update list_{gcs}
else re-send $\text{list}_{gcs}[1]$ at the beginning of the next h-cycle

Fig. 6. The flows of a stream authenticated heartbeat.

there are no interruptions (passive adversary), the GCS sends the number $list_{gcs}[1]$ and the drone responds with the number $list_d[4]$, and both parties update their lists. Lists are updated by shifting numbers two places after two additional draws, n_{2j+3}, n_{2j+4}, from g. That is:

$$\text{list}_{gcs} : [n_{2j}, n_{2j+1}, n_{2j+2}] \leftarrow [n_{2j+2}, n_{2j+3}, n_{2j+4}],$$
$$\text{list}_d : [n_{2j-2}, n_{2j-1}, n_{2j}, n_{2j+1}] \leftarrow [n_{2j}, n_{2j+1}, n_{2j+2}, n_{2j+3}].$$

The other numbers are used when the numbers sent are dropped or substituted or replayed. The GCS always sends numbers at the beginning of a heartbeat cycle (h-cycle, 1 Hz), with the drone responding after receiving the number.

Figure 6 shows a typical round of the protocol (we use the convention that numbers sent as n_j are received as n'_j). This procedure is repeated to get a stream of authenticators: $n_0, \ldots, n_{2j+1}, n_{2j+1}, \ldots$ drawn from g (when the adversary is passive and no sent numbers are dropped), that authenticate the GCS and the drone (mutual authentication): n_{2j} authenticates the GCS while n_{2j+1} authenticates the drone. When numbers *are* dropped, substituted or replayed, we get repetitions (GCS always sends a number) or gaps (the drone only responds if a valid number is received). It is easy to see that any two consecutive numbers drawn from g in this stream will authenticate one party to the other: n_{2j}, n_{2j+1} authenticate the drone to the GCS, while n_{2j+1}, n_{2j+2} authenticate the GCS to the drone. Clearly we have anonymity.

For our application a 6-byte pseudo-random number is sufficient (which is the length of a MAVLink 2.0 truncated HMAC). The heartbeat messages require a minimum amount of computation, and can be pre-computed. If necessary, some of the fields of the MAVLink heartbeat frame can be used to identify the drone or the GCS, although this is not needed since the authenticators uniquely identify the parties. In particular the GCS will store lists $(list_{gcs,d_1}, \ldots, list_{gcs,d_k})$ for all drones that it manages, and use these lists to identify the drones from their authenticators n_{2j+1}. This protocol is based on a protocol for RFID applications and is proven secure in the UC Framework (that supports semantic security) [3].

Securing mission messages (CNSSP 28: A, Ba, Bd, C, D, E). C2 messages are sent by the GCS to execute actions involving navigation (MAV_CMD_NAV), DO (auxiliary) functions such as setting the camera distance (MAV_CMD_DO), and conditional commands (MAV_CMD_CONDITION) to delay actions until some condition is met. MAVLink defines a number of MAV_CMD waypoint command types that are recognized by the autopilot module (ArduPilot). This makes it possible to reduce the size of mission messages to essentially "header type" messages (10 byte header + 6 byte payload and 13 byte timestamped HMAC for MAVLink 2.0), and makes possible C2 message authentication in real time.

If privacy is a concern then one should use an *authenticated encryption with associated data* (AEAD) protocol [28]. AES-GCM (Galois Counter Mode) [18] is an AEAD protocol that combines AES-128 counter mode (CTR) encryption with Galois hash authentication (this protocol is recommended by the NIST SP800-38D [20], and used for IPSec encapsulation). Figure 7 shows a simplified version for the case when the MAVLink header and payload are 128-bit blocks. The input to the protocol is: the AES key k, an initialization vector IV, the plaintext P (the MAVLink payload), and the associated data A. The key k, P and A are 128-bit strings while IV is a 64-bit nonce. The output is: the ciphertext C $= E_{nck}$(P), and the authentication tag T, both 128-bit strings, although for our applications we shall use the leftmost 6 bytes T_{48} of T.

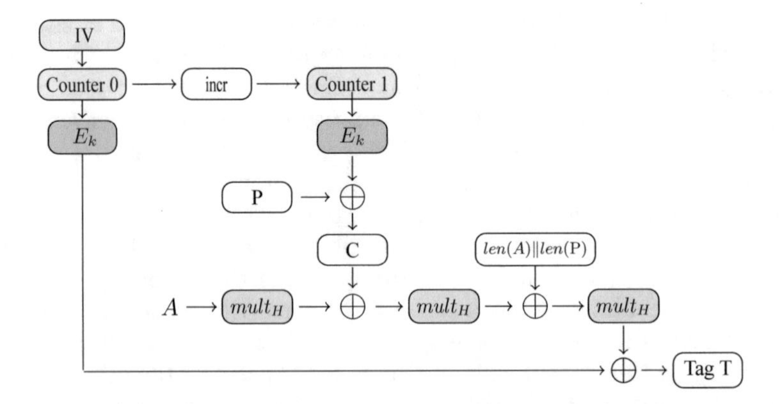

Fig. 7. The GCM authenticated encryption with associated data protocol for a MAVLink mission message for which the header with associated data A and the payload P are 128-bit strings, and IV is an initialization vector. E_k is AES encryption, and $mult_H$ is multiplication in $GF(2^{128})$ by the hash $H = E_k(0^{128})$.

The protocol uses an operator $mult_H$, that is multiplication by the hash key $H = E_k(0^{128})$ in the Galois field $GF(2^{128})$, and a counter whose initial value is $(IV \| 0^{63}1)$. The counter is incremented each time it is invoked. The function $len(X)$ takes a bit string X with length between 0 and 127 and returns a 64-bit string whose value is the length of X, and $X \| Y$ denotes the concatenation of bitstrings X, Y. The operation "$\oplus$" is bit-wise XOR, which is addition of $GF(2^{128})$. For lightweight applications, the time complexity for computing an AES-GCM mission message with a 16-byte payload is 2 µs [18].

The authenticated encryption of a C2 packet consists of the 10-byte header, the 16-byte ciphertext CT with the MAVLink commands and the 6-byte tag T_{48}. The tag authenticates both the header and the ciphertext. The command actions are obtained by decrypting the ciphertext. These operations can also be easily parallelized.

Securing Mission Messages with Captured Data (CNSSP 28: A, Ba, Bd, C). Drones are typically used to capture data with video cameras or sensors, which is then transmitted to the GCS. Such mission messages can be protected using the GCM protocol with input restricted to non-confidential data, that is, with only associated data. This variant is called GMAC, which is an authentication protocol. To authenticate a stream of m 128-bit blocks this protocol requires $(m + 1)$ Galois field multiplications.

Securing Mission Termination Messages (CNSSP 28: A, Ba, Bd, D). Mission termination messages should be short, easy to verify and hard to forge. They also should be timestamped, to prevent "capture and replay" attacks. For such messages we propose to use a basic version of the GMAC variant for which the termination key is the hash key. The mission termination message has a 16-byte header (the 10-byte MAVLink 2.0 header + a 6-byte timestamp) and a 6-byte tag T_{48}. Computing the tag requires only three Galois multiplications. If necessary, mission termination messages can be structured as C2 messages by having encrypted payloads (CNSSP 28: Bc, D). For lightweight applications the time complexity to evaluate such mission termination messages is less than 1 µs [18].

Plans for Recovery and Protection of Classified Materials/Components (CNSSP 28: E). To enforce compliance to a set of rules (a mission planner), an automated V&V approach for continuous (sampling-based) monitoring can be used, as in the Safeguard system [6], or the Automated Mission Planning system [31], with any violations leading to mission termination.

The effectiveness of such an approach depends upon the integrity of components of the UAS and the reliability of the avionics system.

4.4 Bootstrap Loading and Zeroizing (CNSSP 28 Policies: A, Be)

When cryptographic algorithms are used, secret keys have to be generated, managed and distributed. Such keys must be protected so that non-releasable information does not leak if the drone is compromised or captured.

For our Group 1 UAV applications that involve low cost restricted flight-time civilian drones, each drone flight is regarded as a separate session with all keying material zeroized at the end of the session (on power off). To initialize a session, a *secure channel* (not wireless) linking the drone to the GCS and Remote Control is used. The Remote Control (user) selects a fresh random master session key MSK and sends this to the GCS and the drone using the secure channel (if there are several drones each gets a different MSK). Each drone confirms that it received the MSK, and then uses this key to compute its AES-128 key, Galois hash key, a seed for the heartbeat PRNG, and any other keys needed.

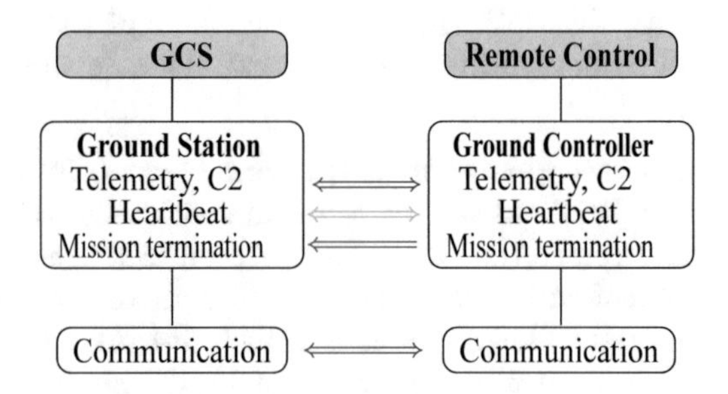

Fig. 8. A CNSSP 28 compliant framework for the GCS and Remote Controller.

4.5 A CNSSP 28 Framework for the GCS and Remote Control

The framework for the GCS and Remote Controller is shown in Fig. 8.

Securing the GCS and Remote Control (CNSSP 28: A, Ba, Bd, C, D, E). The GCS is securely paired with the BM of the UAV and the Remote Controller. Pairing requires mutual device authentication and link key generation [22]). Flight control is established with first-person view (FPV) command and control (C2) instructions (video piloting).

Cross Site Scripting (XSS) attacks can be: *persistent*, *reflective* or *DOM based*. To illustrate such attacks we next give a high level description of a persistent XSS attack which involves a hacker and the client (the drone controller).

1. The hacker finds a weakness in a web application that requires a client to login with a username/password and injects in its server malicious script that steals each visitor's session (authorization) cookie.
2. Each time a client visits the website the malicious script is activated.
3. The session cookie is sent to the client and the hacker.

Defending Against XSS Attacks. As client web browsers evolve, they incorporate an increasingly diverse range of functionalities. At the same time, many common desktop applications extend their functionality to replicate or incorporate the functionality of these browsers. While a security flaw may be an HTML injection, and more specifically XSS, the opportunities for attackers to initiate attacks by exploiting system vulnerabilities grow at an alarming rate. Unfortunately, the delivery methods are becoming so diverse that no single security solution can prevent them. Below we list some approaches commonly used to protect UASs from XSS attacks.

1. **XSS Whitelist Model** [21].
   One way to prevent XSS attacks is to treat an HTML page as a template with slots for untrusted data. Moving untrusted data from slots in the template is not allowed. This is a white list approach that denies everything that is not specifically allowed. Given the way browsers parse HTML, each slot may have different security rules. When putting untrusted data into slots, these rules must be adhered to. Data in slots is not allowed to break out of a slot into a context that allows code execution. The OWASP Prevention Cheat Sheet list [23] proposes seven security rules that prevent the insertion of untrusted data inside normal HTML tags.
2. **OWASP Enterprise Security API** [24].
   This is an open source security web application interface, that works on a whitelist using: (a) a set of security control interfaces, (b) a reference implementation for each security control, and (c) provision for customized implementation.
3. **HTML Defense (sanitizing) Filters**.
   This involves an HTML encoding application in which key characters needed to deliver an XSS attack are encoded. Unfortunately a determined XSS attacker can bypass such filters.
4. **Attack Code discrimination**.
   This includes three technologies: (*a*) BEEP, that only allows Javascript blocks in a whitelist to be executed on client browsers; (*b*) Noncespaces, that generates a random XML namespace for each XHTML document requested by the user, and modifies all trusted XHTML tags in the document with symbols in the namespace; and (*c*) Moving Target Defense, that adds a random attribute to each unsafe element in the web application to distinguish between Javascript code in web applications and injected Javascript code.

Multiple solutions have been proposed for XSS defense, but hackers keep finding new exploits. The situation is aggravated by the complexity of most web application technologies. There is no simple solution, and ultimately the only protection is to try to detect and/or prevent such attacks and use technologies that minimize the risk. The OWASP XSS Cheat Sheet Series [23] lists a number of measures for primary defense (such as turning off the http trace support). Technologies to minimize the risk include XSS protection technologies such as the ones described above (e.g., the OAWSP Enterprise Security API), as well as Docker container virtualization (Sect. 3.4).

4.6 Offensive Security for C2 Messages (CNSSP 28: C, D)

Wyner's wiretap channel model [37] uses an information theoretic (keyless) approach that exploits properties of the wireless medium such as the inherent noise or the superposition property of the medium (interference), to secure communication (jam-based security).

It employs two discrete memoryless channels modeled as stochastic encoders: (a) the main channel with input the transmitted signal X and output the receiver's signal Y, and (b) a wiretap channel with input X and output the eavesdropper's signal Z, a noisy version of X. The capacity of the receiver's channel is $C_{\text{rec}} = \max_{p(x)} H(X|Y)$ (the conditional entropy $-\sum_i p(x_i|y_i)\log_2 p(x_i|y_i)$, maximized over all distributions of X), while the capacity of the eavesdropper is degraded by the noise: $C_{eav} = \max_{p(x)} H(X|Z)$. The secrecy capacity is: $C_s = C_{rec} - C_{eav}$ (the maximum achievable rate at which data can be transmitted secretly).

For our applications we use a computational (secret key) variant of Wyner's model (based on a recent publication for RFID ownership transfer [19]), in which $m \geq 1$ noisy drones act as interferers to degrade signals and mitigate eavesdropping at the physical layer. The drones share a PRNG with the GCS (seed and most recent state) that is used to generate the "noise" (pseudo-random numbers). The numbers are transmitted by the drones and combined with the transmitted signal X, using the fundamental property of superposition of the wireless medium—see Fig. 9. For this model there is only one (noisy) channel, with inputs: the transmitted signal X and m pseudo-random signals $R_1, \ldots R_m$ generated by the shared PRNGs, and output: the signal $Y = f(X, R_1, \ldots, R_m)$, where f is the superposition function.

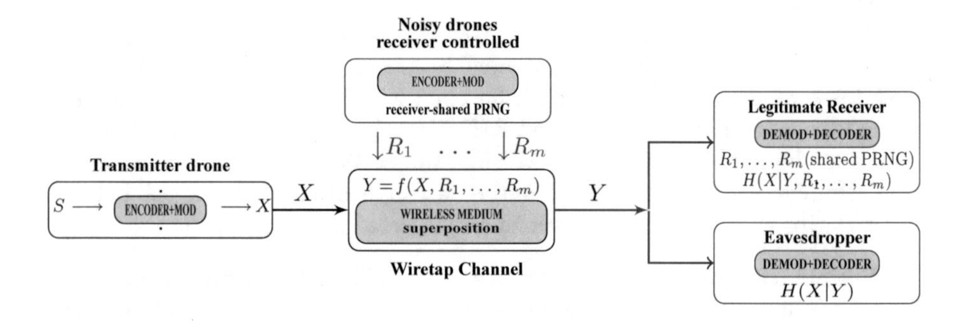

Fig. 9. Wyner's wiretap channel with noisy drones

Figure 10a shows a simple example involving a transmitter drone D_0 (potentially compromised), two noisy drones D_1, D_2 (friend), the eavesdropping ground control station GCS1 (foe), and the intended ground control station GCS2 (friend). D_0 transmits a message S, coded as X, while at the same time D_1, D_2 transmit the coded pseudorandom signals R_1, R_2 generated by the shared (with GCS2) PRNGs. Both GCS1 and GCS2 receive the noisy signal Y. Y is input

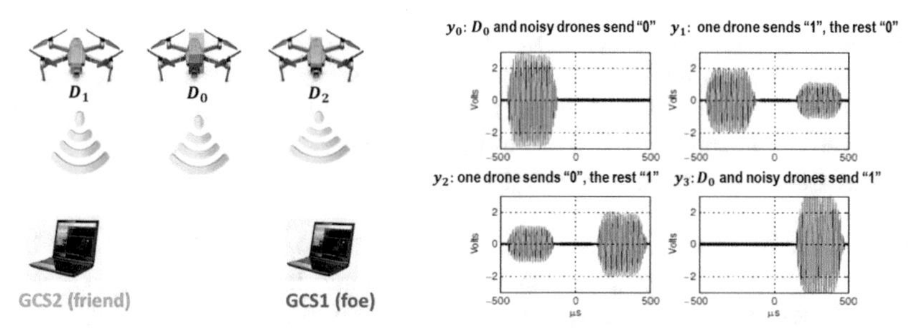

a. Drone D_0 and trusted drones D_1, D_2 monitored by GCS1 (foe) and GCS2 (friend).

b. Superpositioned signals y_0, y_1, y_2, y_3 of D_0, D_1, D_2 (Pulse Position Modulation).

Fig. 10. Drones D_1, D_2 (friend) add noise to the signals of D_0 in a such a way that ground control GCS2 (friend) can remove it, while GCS1 (foe) cannot.

to a maximum a posteriori probability estimator MAP1 of GCS1 and MAP2 of GCS2. If we assume that the wireless medium is noiseless, then the estimate of the intended receiver GCS2 is correct while the estimate of the eavesdropper GCS1 is degraded by the wiretap channel. More specifically, the capacity of the intended receiver's channel is $C_{rec} = \max_{p(x)} H(X|Y, R_1, R_2) = \max_{p(x)} H(X)$, while the capacity of the eavesdropper's channel is $C_{eav} = \max_{p(x)} H(X|Y)$.

Figure 10b shows an example that uses Pulse Position Modulation (PPM) to superimpose the signals transmitted by D_1, D_2. A bit is coded by transmitting a pulse in one of two possible slots (high-low for bit 1, low-high for bit 0). We see that signals y_0, y_3 leak the bit that G_0 transmits to GCS2 (bit 0 and bit 1 respectively), while signals y_2, y_3 hide the bit. It can be shown that for this application with 3 transmitted noisy signals R_1, R_2, R_3,[2] the secrecy capacity is $C_s = C_{rec} - C_{eav} = 0.78$ [19].

Note that the noisy drones wiretap channel can be used both as an offensive tool to obfuscate signals of adversarial drones from their ground control stations, and as a defensive tool to protect C2 and mission terminating channels.

5 Future Work

- **SITL and HITL testing.** Conduct feasibility tests using Software-In-The-Loop simulations as well as Hardware-In-The-Loop simulated tests involving civilian UAV components (e.g. DJI Mavic).
- **Challenges of securing cellular-connected UAVs.** Traditional UAVs rely on line-of-sight (LoS) communication over unlicensed spectrum (the 2.4 GHz and 5.8 GHz band), which has low data rate and is unreliable, insecure and prone to interference. There is significant interest in integrating UASs into 5G cellular networks for reliable and seamless connectivity or, alternatively, to enhance 5G communication capability (e.g. by avoiding coverage holes).

[2] One noisy drone can transmit several noisy signals.

- **Challenges of securing UAV assisted wireless communications.** Dedicated UAVs can be deployed as aerial base stations, access points, or relays to assist 5G terrestrial wireless communications from the sky, and provide UAV-assisted communications. Such applications can facilitate on-demand deployment, network reconfiguration flexibility, and LoS communication links.

We conclude by observing that while novel technologies such as 5G and beyond will expand significantly the functionality and services provided by UASs, they will also introduce new vulnerabilities and new security challenges [9].

Acknowledgments. This material is based upon work supported by the Naval Engineering Education Consortium (NEEC) Award N00174-19-1-0006.

References

1. Arthur, M.P.: Detecting signal spoofing and jamming attacks in UAV networks using a lightweight ids. In: 2019 International Conference on Computer, Information and Telecommunication Systems (CITS), pp. 1–5. IEEE (2019)
2. Autopilot, A.O.S. http://ardupilot.org/
3. Burmester, M., Munilla, J.: Lightweight RFID authentication with forward and backward security. ACM Trans. Inf. Syst. Secur. (TISSEC) **14**(1), 1–26 (2011)
4. CNSSP No. 28, Cybersecurity of Unmanned National Security Systems. http://www.cnss.gov/CNSS/issuances/Policies.cfm
5. Davanian, A., Massacci, F., Allodi, L.: Diversity: A Poor Man's Solution to Drone Takeover. In: PECCS, pp. 25–34 (2017)
6. Dill, E.T., Hayhurst, K.J., Young, S.D., Narkawicz, A.J.: UAS hazard mitigation through assured compliance with conformance criteria. In: 2018 AIAA Information Systems-AIAA Infotech@ Aerospace, p. 1218 (2018)
7. Docker Docs: AC-1 Access Control Policy and Procedures. https://docs.docker.com/compliance/reference/800-53/ac/
8. Dolev, D., Yao, A.: On the security of public key protocols. IEEE Trans. Inf. Theory **29**(2), 198–208 (1983)
9. Fang, D., Qian, Y., Hu, R.Q.: Security for 5g mobile wireless networks. IEEE Access **6**, 4850–4874 (2017)
10. Fuentes, G.: USNI News, June 19, 2018, Pentagon grounds marines' "Eyes in the Sky" drones over cyber security concerns. http://news.usni.org/2018/06/18/pentagon-grounds-marines-eyes-sky-drones-cyber-security-concerns
11. Güvenç, İ., Ozdemir, O., Yapici, Y., Mehrpouyan, H., Matolak, D.: Detection, localization, and tracking of unauthorized UAS and jammers. In: 2017 IEEE/AIAA 36th Digital Avionics Systems Conference (DASC), pp. 1–10. IEEE (2017)
12. Krishna, C.L., Murphy, R.R.: A review on cybersecurity vulnerabilities for unmanned aerial vehicles. In: 2017 IEEE International Symposium on Safety, Security and Rescue Robotics (SSRR), pp. 194–199. IEEE (2017)
13. Kumar, A., Saxena, N., Tsudik, G., Uzun, E.: Caveat EPTOR: a comparative study of secure device pairing methods. In: 2009 IEEE International Conference on Pervasive Computing and Communications, pp. 1–10. IEEE (2009)
14. MAVLink Common Message Set. https://mavlink.io/en/messages/common.html
15. MAVLink Developer Guide. https://mavlink.io

16. MAVLink2. https://mavlink.io/en/guide/mavlink_2.html
17. MAVROS, PX4 Development Guide: the MAVROS ROS package. https://dev.px4.io/v1.9.0/en/ros/mavros_installation.html
18. McGrew, D.A., Viega, J.: The security and performance of the Galois/Counter Mode (GCM) of operation. In: International Conference on Cryptology in India, pp. 343–355. Springer (2004)
19. Munilla, J., Burmester, M., Peinado, A., Yang, G., Susilo, W.: RFID ownership transfer with positive secrecy capacity channels. Sensors **17**(1), 53 (2017)
20. NIST SP 800-38D: Recommendation for Block Cipher Modes of Operation: Galois/Counter Mode (GCM) and GMAC, November 2007. https://csrc.nist.gov/publications/detail/sp/800-38d/final
21. NIST SP 8000-137: Information Security Continuous Monitoring (ISCM) for Federal Information Systems and Organizations. https://nvlpubs.nist.gov/nistpubs/Legacy/SP/nistspecialpublication800-137.pdf
22. NIST Special Publication 800-121: Revision 2, guide to Bluetooth security. https://nvlpubs.nist.gov/nistpubs/Legacy/SP/nistspecialpublication800-121r1.pdf
23. OWASP Cheat Sheet Series. https://www.owasp.org/index.php/OWASPCheat_SheetSeries
24. OWASP Enterprise Security API. https://www.owasp.org/index.php/Category:OWASP_Enterprise_Security_API
25. Pearce, M., Zeadally, S., Hunt, R.: Virtualization: issues, security threats, and solutions. ACM Comput. Surv. (CSUR) **45**(2), 1–39 (2013)
26. Shakeri, R., Al-Garadi, M.A., Badawy, A., Mohamed, A., Khattab, T., Al-Ali, A.K., Harras, K.A., Guizani, M.: Design challenges of multi-UAV systems in cyber-physical applications: a comprehensive survey and future directions. IEEE Commun. Surv. Tutor. **21**(4), 3340–3385 (2019)
27. Shepardson, D.: Reuters, May 20, 2019, DHS warns of data threat from Chinese made drones. http://www.reuters.com/article/us-usa-drones-china/dhs-warns-of-data-threat-from-chinese-made-drones-idUSKCN1SQ1ZY
28. sMAVLink, Secure MAVLink. https://docs.google.com/document/d/1upZ_KnEgK3Hk1j0DfSHl9AdKFMoSqkAQVeK8LsngvEU/edit
29. Srinivas, J., Das, A.K., Kumar, N., Rodrigues, J.J.: TCALAS: temporal credential-based anonymous lightweight authentication scheme for internet of drones environment. IEEE Trans. Veh. Technol. **68**(7), 6903–6916 (2019)
30. Puko, K.F.T.: The Wall Street Journal, October 30, 2019, Interior Department Grounds Aerial Drone Fleet, Citing Risk From Chinese Manufacturers. http://www.wsj.com/articles/interior-dept-grounds-aerial-drone-fleet-citing-risk-from-chinese-manufacturers-11572473703
31. Torens, C., Adolf, F.: Automated Verification and Validation of an Onboard Mission Planning and Execution System for UAVs. In: AIAA Infotech@ Aerospace (I@ A) Conference, p. 4564 (2013)
32. Tridgell, A., Meier, L.: Mavlink 2.0 packet signing proposal (2015). https://docs.google.com/document/d/1ETle6qQRcaNWAmpG2wz0oOpFKSF_bcTmYMQvtTGI8ns/edit#heading=h.r1r08t7lr2pc
33. Ubuntu Server 18.04.3 LTS. https://ubuntu.com/download/server
34. Ward, A.E.: The legal, technical, and practical challenges of countering the commercial drone threat to national security. Technical report, Naval Postgraduate School Monterey United States (2019)
35. What is a Container: Docker, Inc. http://docker.com. Accessed 30 Oct 2019

36. Won, J., Bertino, E.: Securing mobile data collectors by integrating software attestation and encrypted data repositories. In: 2018 IEEE 4th International Conference on Collaboration and Internet Computing (CIC), pp. 26–35. IEEE (2018)
37. Wyner, A.D.: The wire-tap channel. Bell Syst. Tech. J. **54**(8), 1355–1387 (1975)
38. Yasrab, R.: Mitigating docker security issues. arXiv preprint arXiv:1804.05039 (2018)

Using Least-Significant Bit and Random Pixel Encoding with Encryption for Image Steganography

Tapan Soni(✉), Richard Baird(✉), Andrea Lobo(✉), and Vahid Heydari(✉)

Rowan University, Glassboro, NJ 08028, USA
{sonit9,bairdr8}@students.rowan.edu, {lobo,heydari}@rowan.edu
https://cybersecurity.rowan.edu/

Abstract. Steganography is the process of hiding data inside audio, video, and image sources using various encoding methods. This paper presents a novel steganography algorithm that hides files in images using a combination of strong symmetric AES encryption, random pixel selection, and encoding in least-significant bits. The algorithm is implemented in Python 3 and made publicly available as part of a steganography tool, BPStegano. An evaluation using five steganalysis techniques shows that the new algorithm is effective and secure.

Keywords: Steganography · Least-significant bit · AES · Data hiding · Cybersecurity · Random number generator · Python

1 Introduction

Steganography is the process of hiding information inside other sources of information like documents, audio, videos, and images [1] using various different embedding techniques. One such technique of hiding data inside images is called Least-Significant Bit (LSB) data hiding. It uses a 24-bit image where each pixel is made up of three 8-bit values; R, G, and B (Red, Green, and Blue). By changing the LSB of each of the RGB values to the desired binary value, the color of the pixel is changing, but only very slightly. So slightly, in fact, that the human eye cannot differentiate between the original and the modified image. Figure 1 shows how binary data, for example "111", can be hidden using the least significant bit of the RGB value. The encoded binary value and the resulting decimal value is not much different than the original image's decimal and binary value, except that it contains the binary data in the LSB.

By using the LSB data hiding technique, we are relying on the fact that human eyes cannot detect the difference in color from the original image. Although the hash of each image is going to be different, without the original, it is difficult to retrieve the hidden data encoded in the image, especially when other techniques are added such as random pixel encoding and data encryption. In this paper, we propose a steganography tool, BPStegano, which uses the LSB

K.-K. R. Choo et al. (Eds.): NCS 2020, AISC 1271, pp. 139–153, 2021.
https://doi.org/10.1007/978-3-030-58703-1_9

Pixel Value	Decimal Value	Binary Value	Encoded Binary Value	Encoded Decimal Value
Red	255	11111111	11111111	255
Green	240	11110000	11110001	241
Blue	212	11010100	11010101	213

Fig. 1. Encoding "111" inside 1 pixel using RGB LSB data hiding.

data hiding method combined with AES encryption and random pixel hiding to build a robust and secure data hiding program in Python 3 which can hide plain strings and multiple files of any type inside PNG images. The rest of this paper is organized as follows. Over the next section, an overview of the related work is provided. Then, the design of BPStegano is presented. After that, testing results and analysis are presented. Finally, we offer some conclusions and discuss our future work.

2 Related Work

Muyco and Hernandez [2] propose an idea for image steganography using the modified hash-based LSB algorithm [3,4]. By using a modified hash-based LSB algorithm, the encryption and decryption process does not affect the image pixels like the regular LSB encoding method. Although the file sizes are different for both the encoded and original image, histogram and hamming distance steganalysis interprets the original and encoded files to be the same.

Joshi, Gill, and Yadav [5] propose a data hiding technique inside images using the 7th bit of a pixel. The algorithm works by applying a mathematical function to the 7th bit of a pixel value. The pixel and pixel + 1's 7th bits are extracted, and on the basis of a combination of these two values, 2 bits of the secret message can be extracted from each pixel. There can only be four possible combinations of the bits: 00, 01, 10, and 11 and at most, the change in the pixels is going to be +2 or −2. This technique requires both the sender and receiver to know the length of the secret message.

Al-Husainy and Uliyan [6] propose a secret key steganography technique to hide the secret message in the cover image. This technique uses the extraction of chains of elements from the secret key and the use of different key blocks for each block of pixels in the cover image to hide data inside the cover image. This algorithm relies on a large secret key, of at least 256 bytes, and it disperses the data throughout the cover image by utilizing all three RGB values of the pixel.

Kaur and Kocchar [7] implement a steganography technique which used the Discrete Cosine Transform (DCT) to spread the location of the modified pixels over part of the cover image.

Abed et al. [8] implement a method of hiding data inside a video file which uses AES encryption, random video frame selection, and a hardware FPGA steganographic processor to hide the data. A hardware-based approach is utilized because hardware is faster than software in this implementation. The implementation uses a secret key as a seed for a random number generator which is used to select video frames randomly for hiding data. Then, it encrypts the secret data using AES and sends both the selected frames and the encrypted data to the FPGA processor for encoding. The FPGA processor uses the LSB data encoding method to hide the encrypted data inside every pixel of the randomly selected video frames.

Alia et al. [9] propose an enhanced video steganography algorithm based on an exact data matching technique and a random key-dependent data process (EMA-RKDD). Unlike other methods where the cover image or video is modified to include the hidden data, this implementation builds a steganography key which identifies how to decode the specified data from the video by finding values in the video frames that match the secret message. Therefore, in the decoding process, only the steganography key and the original video are needed to extract the hidden data "encoded" inside the video. Additionally, since there is no modification of the cover video, the length of the secret message is theoretically unlimited.

Islam, Modi, and Gupta [10] propose an implementation of image steganography using image edges. It hides data inside the edges of the cover image, where the edges are dynamically selected based on the length of the secret message that is to be hidden. It uses a 2-bit LSB encoding method which decreases the total number of pixels modified. Additionally, since the data is encoded inside the two least significant bits of the cover image, visual detection is almost impossible to deploy, and hiding data in the edges of the image does not produce any visual distortion in the modified images.

Mukherjee, Roy, and Sanyal [11] propose an image steganography algorithm which uses the Arnold transformations and the Mid Position Value (MPV) technique. To hide the data, the cover image undergoes an Arnold transformation which results in a chaotic representation of the original cover image. Then, the data is hidden using the MPV technique. After that, the image goes through an inverse Arnold transformation to create the modified image. The application of the Arnold transformation adds a layer of security as the original pixel positions are scrambled before the secret data is encoded.

Zin and Soe [12] implement similar approaches to image steganography, as described in this paper. In each of the three approaches, the secret data to be hidden is encrypted using RC4 encryption before being sent to the steganography encoder. The first approach uses a basic LSB approach where each pixel had data hidden in its LSB. The second approach uses the Blum Blum Shub pseudo-random number generator to encode data into any pixel that matched to a 1 generated by the number generator. The third approach is similar to the second approach in that the data is hidden based on a random number generated, but the file is divided into a number of blocks and each block has data hidden inside its pixels.

3 Design of BPStegano

For BPStegano, several design requirements are considered. The data to be hidden in the image has to be encrypted using a strong symmetric encryption scheme and it has to be randomly hidden to prevent LSB decoders from detecting the hidden data. To meet those requirements, the NSA approved and highly secure AES encryption cipher is used alongside a random pixel hiding algorithm based on a secret key and the cover image size. The data is hidden inside PNG images to avoid data degradation. BPStegano is able to hide plain text strings and any type and amount of files, limited only by the size of the cover image.

3.1 Encoding Algorithm

Figure 2 describes the encoding algorithm that is designed and implemented for BPStegano. BPStegano takes in four inputs from the user: 1) Data (plain strings or file(s) of any type), 2) Secret key for the AES encryption and random number generation, 3) Source PNG image, and 4) Destination PNG image name. The secret key is fed into the SHA-256 hashing algorithm which outputs a 256-bit hash. This 256-bit hash is used for three purposes: 1) The first 128 bits of the hash are used for the initialization vector for AES, 2) The second 128 bits of the hash are used as the secret key for AES (Note: AES-128 is being used to encrypt the data), and 3) The entire 256-bit hash is used as the seed for the random number generator. Without the secret key, retrieving the hidden data is not possible. It is imperative for the receiving party to know the secret key in order to retrieve the hidden data inside it. The random number generator takes in another input, the total number of encodable bits, which is calculated by multiplying the length of the image by the width of the image and subtracting 1 row (the length) from the product. This is necessary for creating a random number (the hop number) of the correct size, where each pixel needs to be represented by one bit of the random number (Note: These random numbers are extremely large). For example, if the total number of encodable pixels is 100, a 100-bit random number would be generated using the 256-bit hash of the secret key as a seed for the random number generator. The first row (row 0) is subtracted from the total number of pixels in the image to get the total number of encodable pixels because the first row is used for the header.

The header comprises two values: 1) The first 10 pixels of the image contain the length of the ciphertext in binary encoded in the LSB of the RGB values and 2) The zip file header (if the user-specified a file or files as the data to hide). The length of the ciphertext is necessary because BPStegano needs to know where to stop the encoding and decoding process. The zip header is also necessary

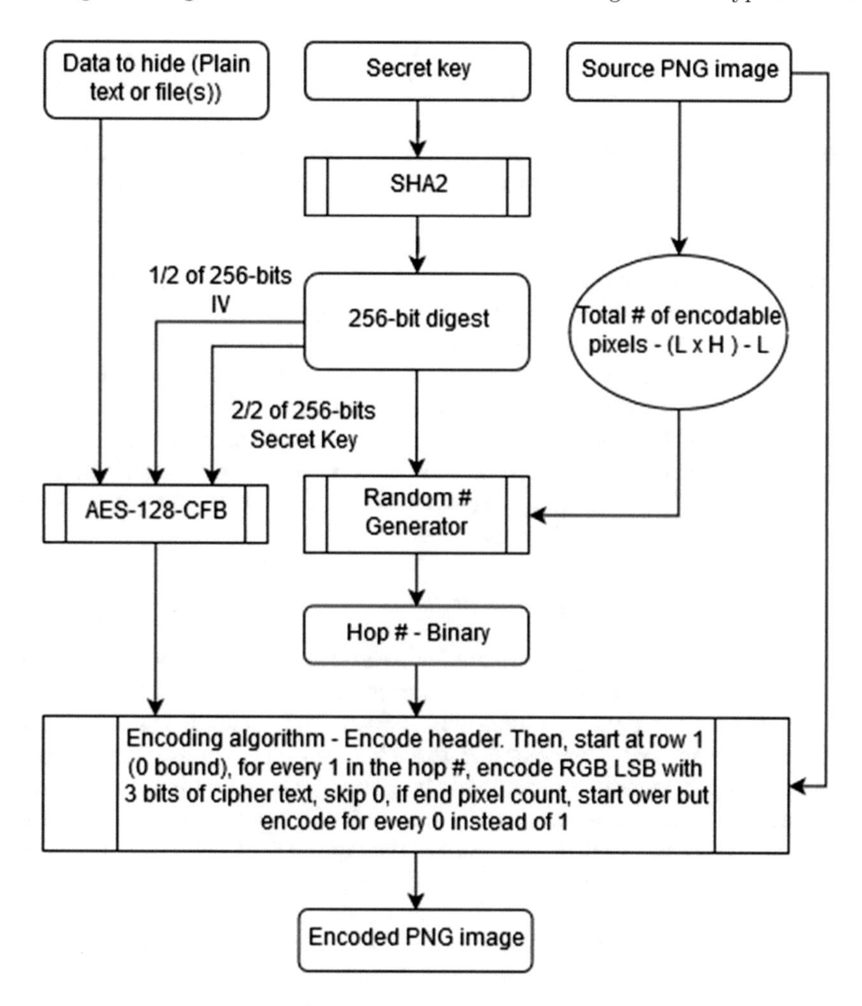

Fig. 2. Encoding algorithm for BPStegano.

because BPStegano creates a zip archive where all the user's files are stored, so only one type of header (zip) has to be processed instead of the hundreds of different types of file headers that exist. This makes BPStegano very robust because it can hide any type of file (.exe, .pdf, .jpg, .xlsx, .docx, .gif, etc.) and more than one file can be hidden in the image. Additionally, since a zip archive is being used, BPStegano works on any platform which supports Python 3 and zip archives which makes it very versatile. Once BPStegano has created the header inside the image, the actual encoding of the ciphertext can begin.

The random number, now converted into binary, is the key to hiding the encrypted data randomly. Starting at the first pixel in row 1 (the image is represented as a 2-D array which is 0 bounded for both the rows and columns) and working through the image, BPStegano encodes 3 bits of ciphertext in each pixel that maps to a 1 in the random number. If a pixel maps to a 0, that pixel is not encoded with any data and is skipped (initially). If BPStegano reaches the end of the image, and there is still ciphertext remaining to be encoded, it returns to the beginning of the image and starts encoding data in the pixels it skipped. In other words, each pixel that maps to a 0 in the random number is now encoded with the remaining ciphertext. In this manner, all the pixels are potentially used, as needed by the ciphertext length, and more importantly, the ciphertext is more spread out through the cover image. Just by looking at the LSB of each pixel, it is very difficult, if not impossible, to determine whether or not it is encoded because the data is encrypted using AES, and the original LSB may have been the same as the cipher bit being encoded therefore no change would appear if the modified image was compared to the original image. Once the encoding has finished, the resulting image is saved as per the user-specified name in the current working directory. The encoding functionality has an exhaustive amount of error checking to help the user navigate through the menu oriented procedure and to identify and correct their mistakes.

3.2 Decoding Algorithm

Figure 3 shows the decoding algorithm for BPStegano. In many ways, the decoding procedure is just the opposite of the encoding algorithm. BPStegano takes in two inputs for the decoding function: 1) Secret key which is used as the seed for the random number generator and the AES decryption, and 2) The encoded image which has the hidden data inside it. BPStegano has built-in functionality to automatically recognize whether the encoded image contains plain text or files by reading the second part of the header (the zip header portion). If the second part of the header matches the zip file header, then there is a file or multiple files hidden inside the image. After reading the header for the size of the ciphertext and whether or not the image contains any files, BPStegano repeats what it did for the encoding function, but instead of changing LSB values, it reads them.

The secret key is hashed using SHA-256, and the 256-bit hash is split into two 128-bit segments, the first one being used as the initialization vector and the second one being used as the secret key for AES. The entire 256-bit hash is then used as the seed for the random number generator which also takes in the total number of encodable pixels as a size parameter for generating the random number. The generated random number, the hop number, is the same number that was generated in the encoding function and is going to be the key to retrieving the hidden data in the correct sequence. By applying the same technique, starting at the first pixel in row 1 (the image is represented as a 2-D array where the rows and columns are 0 bounded) and reading every pixel that maps to a 1 in the random number, BPStegano reads 3 bits of ciphertext at a time. If BPStegano reaches the end of the image and there is still some ciphertext remaining to be found, according to the size of the ciphertext encoded in the first part of the header, then BPStegano returns to the beginning of the image and starts reading pixels that map to a 0 in the random number. Once all of the ciphertext is read, it is sent to the AES decryptor. If the header of the

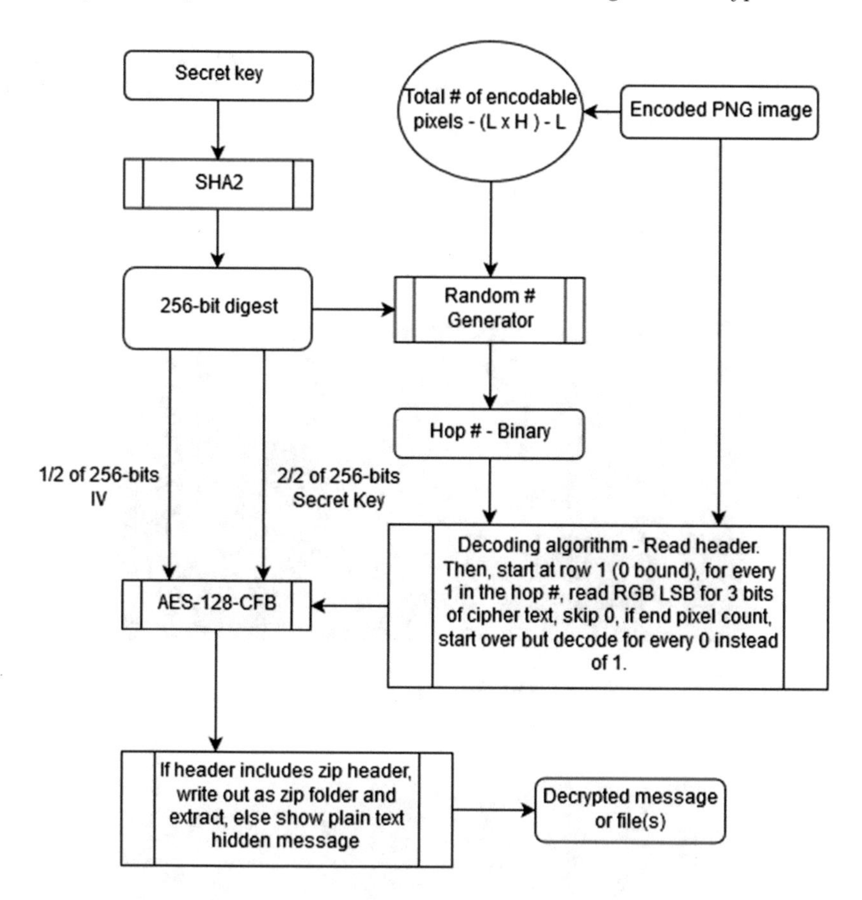

Fig. 3. Decoding algorithm for BPStegano.

image included the zip file header, then BPStegano writes out the decrypted bytes as a zip archive and then extracts the zip archive into a HIDDEN_DATA folder. BPStegano includes the functionality where if a HIDDEN_DATA folder already exists (from previous decoding sessions or if the user put one there), then it will ask the user if they want to overwrite or append to that folder for safety reasons. If the header does not include a zip file header, then BPStegano knows that the hidden data was a plain string and displays it appropriately. The decoding algorithm, as with the encoding algorithm, includes a fair amount of error checking to help the user and to alert them if the secret key is incorrect or the encoded source image does not contain any data.

4 Results and Analysis

For testing and analysis, two images are selected, truck.png [13] and shuttle.png [14]. Two different types of input are tested, a plain text string, and several files. The plain text string used was 99 paragraphs of Lorem Ipsum text which is about 18 pages of text containing 718 lines. For the files, a gif, a text file, and an executable were hidden. The analysis of BPStegano using five steganalysis techniques follows in Sects. 4.1 through 4.5

4.1 Visual Comparison

Fig. 4. Original image - truck.png

Fig. 5. Modified image which contains a hidden string - plain_text.png

Figures 4 and 5 look exactly the same except that Fig. 5 contains text that is encoded randomly into the LSB of each pixel.

4.2 SHA-256 Image Hash Values

The SHA-256 hashing algorithm was used to calculate the hash value for each image set, truck.png vs. plain_text.png and shuttle.png vs. files.png. As predicted, each image in the set had a different hash value because the modified image contained hidden data whereas the original image was not altered.

4.3 Software Based Image Comparison

Because LSB data encoding only encodes data into the rightmost bit of the RGB binary value, it changes the color of the pixel only slightly. Even though a single changed bit is enough the change the hash of the image completely, as discussed in the previous section, to the human eye the changes are impossible to detect. Software is available to find the differences between two images.

Fig. 6. imagonline.com results.

Figure 6 is the result of another image comparison tool [15] which compares at a fuzz percent of zero. Fuzz describes how exact the comparison between the two images is [16]. If the fuzz was set to 0% (the lowest possible value), then every single RGB value in every pixel would be checked in both images and even the smallest difference would be highlighted [16]. Imgonline uses red dots to identify the differences in the images to show where the data is encoded. The dots stop about two thirds through the image because BPStegano has completed the encoding of the data. As seen in the figure, there is an even distribution of the data throughout the encoding area.

Figure 7 shows a comparison between the files.png and the shuttle.png images using an online image difference finder [17]. Even when the fuzz is set to an extreme value of 1%, there are no differences detected between the two images. Only a fuzz level of 0%, as shown earlier, shows the differences in both the images.

By hiding encrypted data into the LSB of each RGB value and using a random pixel encoding scheme, BPStegano is able to hide plain strings and any amount (limited by the size of the source image) of files of any type inside an image while retaining the visual similarity and integrity of the original image.

Fig. 7. Online-image-comparison.com results.

4.4 Mean Square Error and Peak Signal to Noise Ratio

Mean Square Error (MSE) is the square of the errors between the cover image and the modified image [7]. If the MSE is low, then there are a very small amount of errors, and if the MSE is high, then there are a large amount of errors. Equation 1 is used to calculate the MSE where the H and W are the height and width of the image, respectively, P(i, j) represents the original cover image, and S(i, j) represents the modified image [4]:

$$MSE = \frac{1}{H * W} \sum_{i=1}^{H} (P(i,j) - S(i,j))^2 \qquad (1)$$

The MSE calculated for truck.png and plain_text.png was a miniscule 0.0204. That means that there is a very small difference between the two images that can be determined by a quantitative analysis. The MSE calculated for the shuttle.png and files.png was a relatively small amount of 0.1109. Considering that there was more data hidden in the files.png image, 0.1109 is a small value to have because relatively speaking, it shows that the modified image is very similar to the original image, especially for having 1,928,920 pixels encoded with data out of 6,108,000 pixels compared to the 43,682 pixels encoded for plain_text.png.

Peak Signal to Noise Ratio (PSNR), Eq. 2, is used as a quality measurement to compute the signal-to-noise ratio in decibels between two images [4,18]:

$$PSNR = 10log_{10}\frac{L^2}{MSE} \tag{2}$$

The higher the PSNR value, the better the quality of the modified image [18] since the noise (difference) between the pixels in the images is low. The PSNR value for the shuttle.png and files.png was calculated to be 57.68. This is a relatively high value considering that visually the images look exactly the same, but as shown in Fig. 6, the top 66% of the image has data hidden in it. For truck.png and plain_text.png, the PSNR value calculated was 65.03. This means that the plain_text.png has a very close resemblance to truck.png. The MSE and PSNR values are quantitative metrics used to determine how close modified images are to their original counterparts. By encoding with BPStegano, the modified images are almost identical to the original images.

4.5 Image Histogram Analysis

In steganalysis, histograms are used to show the frequency of pixel intensities in an image [19]. A visual comparison of the images can be performed by creating histograms for images and overlaying them one on top of another to see the change in intensity frequency, since LSB encoding changes the pixels to hide data [19]. Figure 8 shows a zoomed-in histogram which shows some of the largest differences in pixel intensity for truck.png and plain_text.png. The difference in the frequencies for each intensity is very small because the LSB of each pixel only changes the intensity very slightly. For some pixel intensities, truck.png has a greater frequency and for others, plain_text.png has a greater frequency. The differences in pixel intensity depend largely on what the original intensity of the pixel was and if the LSB was changed when the ciphertext was encoded. Figure 9 shows another zoomed-in histogram which shows section with the most differences in pixel intensity for shuttle.png and files.png. Files.png contains almost 45 times more data than plain_text.png resulting in the frequency of

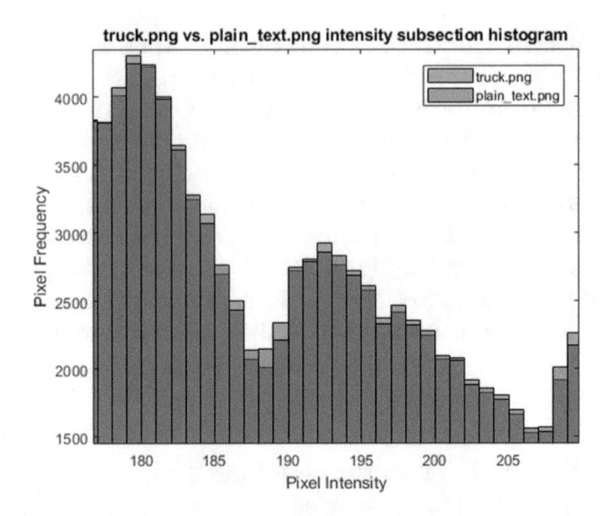

Fig. 8. truck.png vs. plain_text.png zoomed-in intensity histogram

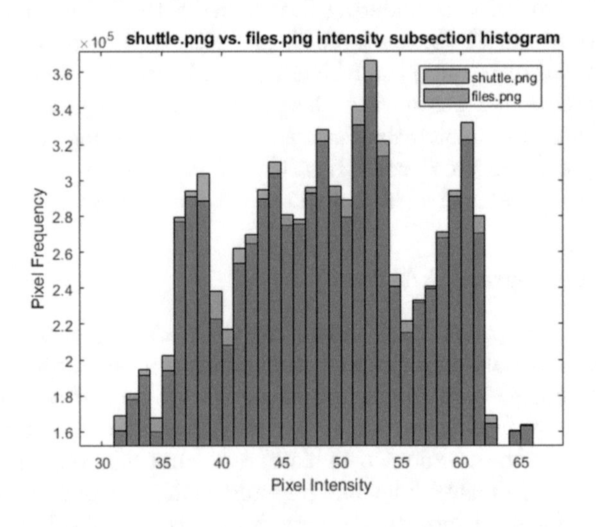

Fig. 9. shuttle.png vs. files.png zoomed-in intensity histogram

differences being more prevalent than the differences in Fig. 8 because more pixels are manipulated. By using the random pixel selection algorithm, BPStegano spreads out the manipulation of pixels. Additionally, because the differences in the images are very small, histograms of the images look very similar as shown in Fig. 10.

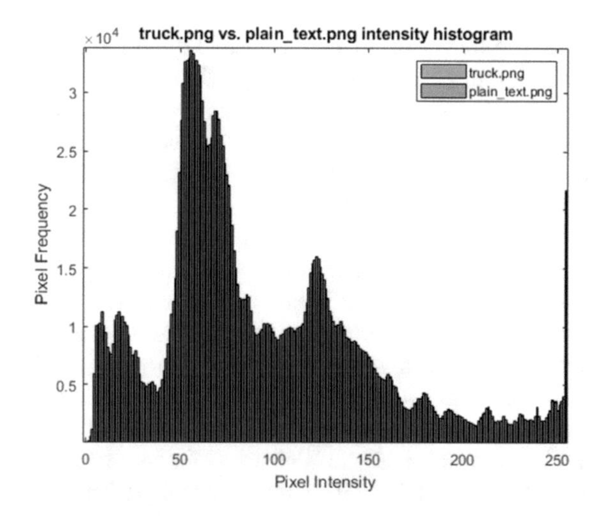

Fig. 10. truck.png vs. plain_text.png intensity histogram

5 Conclusion and Future Work

In this paper, a steganography tool, BPStegano, is described. It uses the LSB data hiding method combined with a random pixel encoding algorithm to distribute the data throughout the image. Additionally, AES encryption is deployed against the data to encrypt it before it is hidden inside the image. For 24-bit images, where each pixel has an 8-bit R, G, and B value, 3 bits of data can be stored in the LSB of every pixel. By working with the image as a 2-D array, the image is able to be manipulated using standard array manipulation techniques. BPStegano uses the secret key's SHA-256 hash as a seed for the random number generator, a unique random number is created that is used to hide the encrypted data throughout the image by encoding data at pixels which correspond to a 1 in the random number's binary representation. If the algorithm runs out of image space, i.e., at the end of the image during the initial pass, it repeats the encoding process from the beginning of the image, except this time it encodes pixels that map to a 0 instead of 1 thus utilizing all the pixels in the image. Multiple files of any type are supported by using a zip archive to combine all the files under one file type, so BPStegano only has to deal with one type of file header instead of the literal hundreds of different file types. The decoding function works in a similar way to the encoding function, except instead of manipulating the LSB, it reads them, decrypts them, and then based on the image header, creates a folder to hold all the hidden files or displays the hidden string. Five steganalysis techniques are employed to evaluate the effectiveness of the BPStegano's algorithm including visual analysis, calculating SHA-256 hash differences, using online image difference tools to detect differences between the original and modified image, Mean Square Error and Peak Signal to Noise Ratio quantitative metric analysis, and histogram analysis. In conclusion, BPStegano is a robust and secure steganography tool that can be used to hide any kind of data inside an image using a random LSB pixel encoding scheme.

The most important part of the future work is to add new features, maintain the program, and to fix any bugs that are discovered through regular usage and testing. New features such as different encoding mediums (audio, video, gifs, and network data packets) will be researched, tested, and implemented to allow users to hide greater amounts of data in different mediums. As seen in Fig. 6, the encoding process uses a random pixel encoding algorithm, but it is still sequential. A new algorithm for hiding will be researched and implemented where the data is spread throughout the image. This way, no matter how small the input data is, it will always be distributed throughout the entire image, instead of just the encoding area calculated by the length of the cipher text. Options to disable encryption and select different types of encryption will also be deployed. A GUI application will also be considered in order to show the encoding and decoding procedure in a more visual and aesthetically pleasing manner and to show the before and after images side by side. The code for BPStegano is located on GitHub [20].

References

1. Kaur, H., Rani, J.: A survey on different techniques of steganography. In: MATEC Web of Conferences, vol. 57, p. 02003 (2016). https://doi.org/10.1051/matecconf/20165702003
2. Muyco, S.D., Hernandez, A.A.: A modified hash based least significant bits algorithm for steganography. In: Proceedings of the 2019 4th International Conference on Big Data and Computing, Series ICBDC 2019, pp. 215–220. ACM, New York (2019). https://doi.org/10.1145/3335484.3335514
3. Meshram, A.G., Patil, R.: Secure secret key transfer using modified hash based LSB method. Int. J. Comput. Sci. Inf. Technol. **5**, 3 (2014)
4. Dasgupta, K., Mandal, J.K., Dutta, P.: Hash based least significant bit technique for video steganography HLSB. Int. J. Secur. Priv. Trust Manag. (IJSPTM) **1**(2), 11 (2014)
5. Joshi, K., Gill, S., Yadav, R.: A new method of image steganography using 7th bit of a pixel as indicator by introducing the successive temporary pixel in the gray scale image. J. Comput. Netw. Commun. **2018**, 10. https://www.hindawi.com/journals/jcnc/2018/9475142/
6. Al-Husainy, M., Uliyan, D.: A secret-key image steganography technique using random chain codes. Int. J. Technol. **10**, 731–740 (2019)
7. Kaur, G., Kochhar, A.: A steganography implementation based on LSB & DCT. Int. J. Sci. Emerging Technol. Latest Trends **4**, 7 (2012)
8. Abed, S., Al-Mutairi, M., Al-Watyan, A., Al-Mutairi, O., AlEnizy, W., Al-Noori, A.: An automated security approach of video steganography-based LSB using FPGA implementation. J. Circuits Syst. Comput. **28**(05), 1950083 (2019). https://doi.org/10.1142/S021812661950083X
9. Alia, M.A., Maria, K.A., Alsarayreh, M.A., Maria, E.A., Almanasra, S.: An improved video steganography: using random key-dependent. In: 2019 IEEE Jordan International Joint Conference on Electrical Engineering and Information Technology (JEEIT), pp. 234–237, April 2019

10. Islam, S., Modi, M.R., Gupta, P.: Edge-based image steganography. EURASIP J. Inf. Secur. **2014**(1), 8 (2014)
11. Mukherjee, S., Roy, S., Sanyal, G.: Image steganography using mid position value technique. Proc. Comput. Sci. **132**, 461–468. http://www.sciencedirect.com/science/article/pii/S1877050918308949
12. Zin, W.W., Soe, T.N.: Implementation and analysis of three steganographic approaches. In: 2011 3rd International Conference on Computer Research and Development, vol. 2, pp. 456–460, March 2011
13. Cybertruck. https://www.tesla.com/cybertruck
14. Space shuttle project—NASA image and video library. https://images.nasa.gov/details-8898508
15. imgonline-com-ua. https://www.imgonline.com.ua/eng/difference-between-two-images.php
16. Online image diff - website for easy online image comparison. https://online-image-comparison.com/faq
17. Online image diff - website for easy online image comparison. https://online-image-comparison.com/result
18. Horé, A., Ziou, D.: Image quality metrics: PSNR vs. SSIM. In: 2010 20th International Conference on Pattern Recognition, pp. 2366–2369, August 2010
19. Kaur, H., Sohi, N.: A study for applications of histogram in image enhancement. Int. J. Eng. Sci. **06**, 59–63 (2017)
20. Tapan, S.: TapanSoni/BPStegano. original-date: 2019-12-07T03:23:31Z. https://github.com/TapanSoni/BPStegano

Automated Linux Secure Host Baseline for Real-Time Applications Requiring SCAP Compliance

Zack Kirkendoll(✉), Matthew Lueck, Nathan Hutchins, and Loyd Hook

University of Tulsa, Tulsa, OK 74104, USA
zack-kirkendoll@utulsa.edu

Abstract. With the emergence of Linux real-time applications and the ever-growing cyber security requirements being imposed on those systems in high security environments, a need has developed for an automated and quantitative approach for developing a solution. The development and mandate for a Department of Defense Windows Secure Host Baseline has provided a framework for how to approach an equivalent Linux Secure Host Baseline. However, a method for automating a cyber security compliant Linux distribution with real-time capability has not yet been created.

Utilizing National Institute of Standards and Technology (NIST) approved tools and security guidelines (DISA STIG), this work has produced an automated Linux distribution capable of performing real-time functions with a 96.15% security compliance performance metric. The compliance deficiencies have been addressed to a satisfactory level based on their context. The promising results show that this approach can be utilized by other organizations for Linux real-time application development and deployment with customized environments specific to their needs while maintaining a high level of security posture. The methodology proposed provides a template for designing, customizing, and maintaining the solution.

Keywords: Security · Real-time · Linux · SCAP · STIG · DISA · CentOS

1 Introduction

Industrial, aviation, and safety-critical applications often require hard guarantees for system response times. These real-time applications require operational timing constraints between triggering events and the applications response to that event. This determinism of meeting schedule deadlines is accomplished through a real-time operating system (RTOS). The latency of a task depends on other tasks running at equal or higher priorities.

The Linux Operating System (OS) has grown into the most important operating system in the world and has begun to dominate each market it has entered. Linux has a dominant use within real-time applications such as aviation, flight simulators, medical, military, controls, audio/video, with emerging use in automotive and safety-critical

K.-K. R. Choo et al. (Eds.): NCS 2020, AISC 1271, pp. 154–169, 2021.
https://doi.org/10.1007/978-3-030-58703-1_10

products. As cybersecurity requirements continue to grow, the implementation of compliant security controls become increasingly important. The security control distribution framework should provide an automated and flexible approach for use by organizations of all sizes and use cases.

The primary objective of this work is to provide a means to automate the build and distribution of a Linux Secure Host Baseline (SHB) for use in real-time applications with acceptable Security Content Automation Protocol (SCAP) compliance utilizing freely available tools. A driving factor for this approach is the Department of Defense (DoD) Chief Information Office (CIO) memorandum dictating all DoD components transition to the Windows 10 SHB, focus on strengthening the cyber security posture of those systems, while streamlining the IT operating environment [1].

The system focused within this work utilizes the Community ENTerprise Operating System (CentOS) Linux operating system, the Real-time Linux Kernel Extensions (PREEMPT_RT Patch), Defense Information Systems Agency (DISA) Red Hat Enterprise Linux (RHEL) Security Technical Implementation Guide (STIG), and the SCAP Compliance Checker (SCC) and OpenSCAP tools. Each of these independent approaches will be covered in the following background section.

2 Background

2.1 Security Content Automation Protocol (SCAP)

SCAP is a common method for utilizing a specific set of standards for enabling automated security vulnerability compliance evaluation for organization's systems through quantitative and qualitative metrics. Generally, SCAP is a synthesis of interoperable specifications derived from community driven ideas and participation to ensure a broad range of use cases for a common evolving security goal.

The National Institute of Standards and Technology (NIST) describes SCAP content checklists as adhering "to the SCAP specification in NIST SP 800-126 for documenting security settings in machine-readable standardized SCAP formats. SCAP content checklists can be processed by SCAP-validated products, which have been validated by an accredited independent testing laboratory as conforming to applicable SCAP specifications and requirements" [2].

At a high-level, SCAP compliance is controlled using SCAP security guides (security recommendations) together with SCAP compliance tools to audit systems automatically. Generally, this means running an automated tool to check non-compliance security settings on a given system and then remediate those non-compliance settings iteratively until the system has reached a satisfactory level of compliance.

2.2 Security Technical Implementation Guide (STIG)

A STIG (security guide) is a cybersecurity methodology for standardizing security protocols within a network or computer to enhance the security posture of the entire system through configuration settings. The primary implementation of STIGs occurs on desktop computers or servers to prevent system access from unauthorized users. STIG design considerations typically cover accessibility, networks, routers, firewalls, domains, and switches. STIGs are specifically utilized by the Department of Defense (DoD). The chosen STIG security profile for this work is `xccdf_org.ssgproject.content_profile_stig-rhel7-disa` which applies to RHEL 7 as provided by DISA. This profile will be used to determine the system compliance.

2.3 Government and Industry Collaboration – NIST, DISA, NSA, USGCB

There exists several widely accepted standardized government level security protocols and practices associated with specific agencies. These include NIST 800-53, DISA/STIG, National Security Agency (NSA), and United States Government Configuration Baseline (USGCB) which provides a set of security measures for several operating systems. While NIST is the primary agency for determining SCAP compliance approaches, both DISA and the NSA are recommended by NIST as agency-produced checklist vendors [3]. The primary focus for this work will be on NIST and DISA collaboration.

2.4 Windows 10 Secure Host Baseline (SHB)

DISA and the NSA co-developed and maintain a Microsoft Windows 10 standard desktop framework known as the DoD Windows 10 Secure Host Baseline (SHB). The purpose of the Windows SHB provides an automated and flexible approach for assisting the DoD in deploying the latest release of Windows 10 based on recommended practices, to be consumed by organizations of all sizes ranging from the large enterprise to small deployments. The SHB primarily focuses on heightening the security baseline of the operating system through Group Policy objects, compliance checks, and configuration tools for system administrators.

The framework can be downloaded from DISA and installed using a development installation of Windows 10 with no STIG policies implemented. The framework creates a reference virtual machine and applies STIG policies, and required components. Optional components and applications are installed along with any other organizational required customizations. A reference image (ISO) is then captured from the reference machine and optionally installed to deployment media (USB hard drive or optical media).

The framework includes core components and applications along with optional applications that can be installed during the build and capture phase. The framework also includes settings for the following STIGs by default:

- Microsoft Windows 10
- Microsoft .NET Framework 4
- Internet Explorer 11
- Windows Firewall

2.5 Community ENTerprise Operating System (CentOS)

Community ENTerprise Operating System (CentOS) Linux is a community maintained, stable open source operating system built with changes from the Red Hat Enterprise

Linux Source Code. While being a community project, CentOS Linux does not inherit Red Hat Enterprise Linux certifications or evaluations. However, CentOS is commonly used for Linux applications requiring real-time performance or for system servers. CentOS comparatively has a longer release and support cycle which results in a very stable system.

2.6 Real-Time Linux Kernel Extensions (PREEMPT_RT Patch)

A real-time operating system is needed for running applications that have real-time requirements. Real-time applications are primarily focused on determinism and maximizing response time rather than throughput. Real-time applications have operational deadlines between an event trigger and its response to that event. Scheduling through priority schemes and ensuring higher priority tasks are run prior to lower priority tasks requires preemption. Preemption allows lower priority tasks to be interrupted in favor of higher priority tasks with the intent to resume those tasks later [4].

The PREEMPT_RT patch converts Linux into a fully preemptible kernel. This approach allows the user space to preempt kernel tasks to ensure determinism within the real-time application. Utilizing this patch with Linux provides a low cost and mature solution for a real-time operating system [4].

The official PREEMPT_RT kernel patch is maintained by kernel.org with different versions matching specific versions of the mainline kernel source. The latest versions can be downloaded from The Linux Foundation's Real-Time Linux Project page. The kernel patch can be integrated into a stock Linux distribution in multiple ways including being added to the kernel source tree and built using make, precompiled packages for RHEL for Real Time 7, and CentOS 7 repository maintained by CentOS [5].

Linux provides many features which support viewing, modifying, and tuning real-time performance of systems. One such system that is installed by default is Tuna which provides a graphical user interface for monitoring and configuring of task priorities, scheduling policies, and CPU affinities of tasks and threads in real-time to ensure real-time requirements are being met.

2.7 SCAP Compliance Tools – OpenSCAP (OSCAP), SCAP Compliance Checker (SCC)

SCAP compliance tools are validated scanners capable of authenticated vulnerability scanning available for both DoD and non-DoD use. These tools aid administrators and auditors with assessment, measurement, and enforcement of security baselines that adhere to some security standard. In this case the security standard is maintained by NIST with specific SCAP versioning certification. Generally, the usage and process of using compliance tools is an iterative process of checking the system against the security policies and remediating any discrepancies and vulnerabilities that are discovered. DISA provides benchmarks for use with SCC, but not OSCAP, so SCC is used for determining compliance for this system.

OpenSCAP is a security compliance tool that checks security configuration settings and examines indicators of compromise by comparing the existing operating system settings to the chosen security policies. OpenSCAP is validated by NIST for use on

RHEL, however, these certifications do not directly apply to CentOS even though CentOS is built from the Red Hat source. In this scenario, OpenSCAP is utilized for applying the RHEL 7 DISA STIG security profile to the CentOS 7 operating system during the initial setup process and remediating some of the deficiencies discovered.

The SCC is an automated compliance scanning tool that leverages the DISA STIGs for analyzing and reporting on the security posture of the scanned system [6]. The tool has no license cost to government or contractors, performs compliance scanning using SCAP content, creates an HTML report page as an output, and has graphical and command line interfaces [7].

At a high level, STIGs are just guidelines and during adjudication the examiner of the SCAP compliance results must determine whether non-remediated deficiencies are acceptable as risks for that organization using a more nuanced approach.

3 Related Work

3.1 Security Hardening Technology

The initial basis for generating hardening scripts used within the proposed framework has been expanded on from Frank Caviggia's CentOS 7 project used for a standard distribution that is not capable of real-time performance [8]. Caviggia developed these scripts while working at Red Hat and along with others in the community continues to maintain them. The primary goal of Caviggia's scripts is to configure and harden the baseline CentOS ISO using SCAP Security Guide (SSG). Red Hat maintains a RHEL based fork of Caviggia's project [9]. The security hardening is implemented using RHEL's standard kickstart technology.

3.2 Automated Hardening and Testing Linux

Henttunen presents a method to sufficiently securing a Linux operating system running on a VMWare workstation utilizing automated installation scripts and open source auditing tools [12]. The approach focused on a USGCB security standard compliant hardened environment rather than the DISA/STIG compliance approach. The VMWare environment is applicable to non-real-time and server applications typically used for cloud-based services.

4 Design and Implementation

The goal of the Linux SHB development is to mirror the framework and design approach achieved by the Windows SHB while providing adequate SCAP compliance through quantifiable metrics and enabling real-time capability. This approach will include the necessary real-time patch required for real-time application deployment through use of a custom system profile. The re-spin script and kickstart files were developed and tested against CentOS 7.6 1810. The approach for a re-spin is to modify and provide updates to an existing ISO image. The high-level approach described within this section can be visualized using Fig. 1.

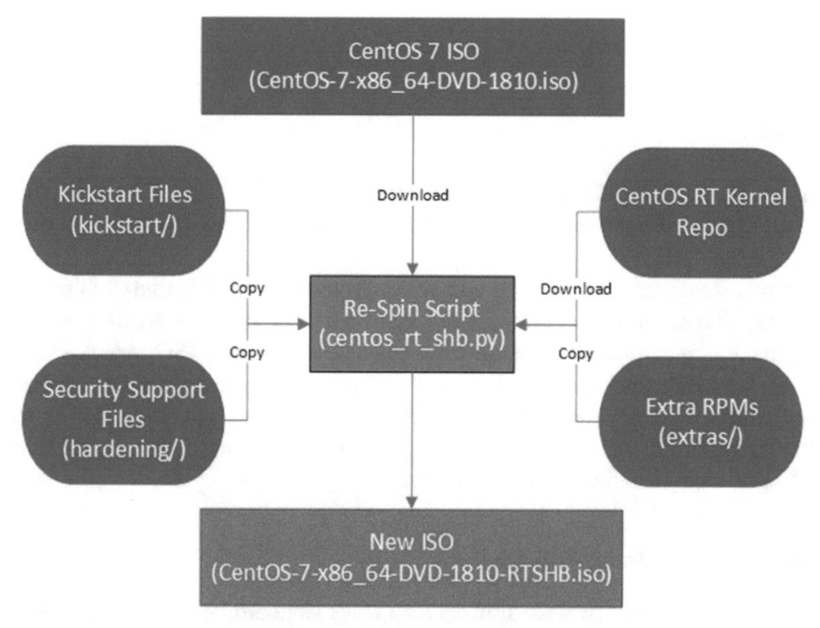

Fig. 1. CentOS ISO re-spin process

4.1 Base Linux Distribution

The base operating system distribution selected is CentOS 7. This specific distribution is freely available, closely matches Red Hat Enterprise Linux releases, and has been widely used in industry. This approach will allow exact STIG configuration items to be applied and the same SCAP compliance tools to be utilized across both systems. CentOS will include the PREEMPT_RT Patch within its installation process.

4.2 Test Setup and Build Environment

The test setup utilizes a virtual machine for ease of use and reproducibility. This approach will not limit functionality and will translate directly for use on real hardware. The virtual machine software used is Oracle VirtualBox 5.2.22. The computer hardware used is a Dell with Intel i7 x64-based processor, 16 GB of RAM, and running Windows 10.

The CentOS re-spin script (centos_rt_shb.py) must be executed on an existing CentOS 7 installation with access to the internet. It is recommended that the build machine has at least 40 GB of free hard drive space and at least 8 GB of RAM. This script can be started within a virtual machine or on real hardware.

4.3 ISO Build Approach

The script is started within an existing CentOS 7 installation and will require the following packages. The script will install these packages using yum if they are not already installed.

- `yum-utils`
- `genisoimage`
- `createrepo`
- `isomd5sum`

The latest CentOS 7 ISO is downloaded directly from the CentOS website through their verified distribution directory. This provides the base Linux distribution used for generating the SHB ISO.

The real-time patch to the kernel will be integrated into the repository of the installation media. The real-time repository is cloned to the local machine and the downloaded packages are copied into the Package directory of the extracted ISO image. The script modified the ISO's package group repository data to include the RT group that was defined within the real-time repository. When the `createrepo` command is run on the new ISO package directory, the RT group is available for installation by the standard CentOS graphical installer and kickstart files.

4.4 Security Hardening Approach

The security hardening approach utilizes Red Hat's standard kickstart technology. The CentOS 7 kickstart project created by Caviggia that is community maintained is the base implementation that is expanded upon and tailored for this approach.

The top level kickstart file is called /kickstart/hardened-centos.cfg. This is executed during the CentOS installation and calls a graphical menu (menu.py) to allow the user to configure some basic attributes of the hardened system.

The following configuration option are available (see Fig. 2).

1. **Hostname:** This field allows the user to set the hostname of the installed system.
2. **System Profile:** This dropdown menu allows the user to select which profile will be installed.
3. **System Classification:** This dropdown menu allows the user to select the classification level of the system. The following options are available.

   a. Unclassified
   b. Unclassified/FOUO
   c. Confidential
   d. Secret
   e. Top Secret
   f. Top Secret/SCI
   g. Top Secret/SCI/NOFORN

   .

4. **SCAP Security Guide Profile:** This dropdown menu allows the user to select the security profile to implement. Currently, DISA STIG is the only option.
5. **System Hardware Information:** This displays information about the hardware of the system.

6. **Available Disks:** These check boxes allow the user to select which disk(s) to install onto.
7. **Encrypt Drives with LUKS:** This checkbox allows the user to select whether LUKS encryption is used on the filesystems.
8. **Disable USB:** This checkbox allows the user to disable USB support.
9. **Lock root:** This checkbox disables the root account. A user account called admin is added with `sudo` capabilities to be used for system administration.
10. **Install Real-Time Kernel:** This checkbox installs the real-time kernel packages. This installs the RT package group. This is part of the expanded capability.
11. **Kernel in FIPS 140-2 Mode:** This checkbox enables FIPS 140–2 encryption on the system.
12. **Required/Optional LVM Partitioning Percentage:** This allows the user to define what percentage of the installation disk should be allotted to each partition. The required partitions are required to be defined per the STIG requirements. The values entered in the fields must add up to 100%.
13. **Network Configuration:** This button opens a dialog window allowing the user to configure the network interfaces of the machine.
14. **Help:** This button opens a dialog window with help information.
15. **OK:** This button is clicked once all other configuration is complete. The user will be prompted to enter a password. The password must be 15 characters long. The entered password will become the password for the root and admin accounts, the grub password, and the key to decrypt the drive is LUKS is enabled.

After the user completes the configuration using the graphical menu, the installation continues without user input.

The kickstart file utilizes the `%pre` and `%post` sections for scripting the security hardening process in its entirety. Selections from the graphical menu that were inputted by the user adds custom code to these sections as appropriate. The script will utilize the OSCAP tool provided within the OpenSCAP packages to apply the RHEL 7 DISA STIG security profile (`xccdf_org.ssgproject.content_profile_stig-rhel7-disa`) during installation.

The security profile itself satisfies approximately 80% of the STIG requirements for the system. Additional scripts and tools (including /hardening/supplemental.sh) complete the remaining remediations. SSH services are installed by default but a user must be a member of the `sshusers` group to log in remotely, which will be managed by the administrator.

4.5 Security Tools and Services

The following tools and services are included to satisfy the STIG requirements and in general to improve the security posture of the total system.

- **Smart Card Support:** Packages to support login services requiring a CAC or other smart card.
- **Auditd:** The standard Linux auditing daemon configured per STIG requirements.

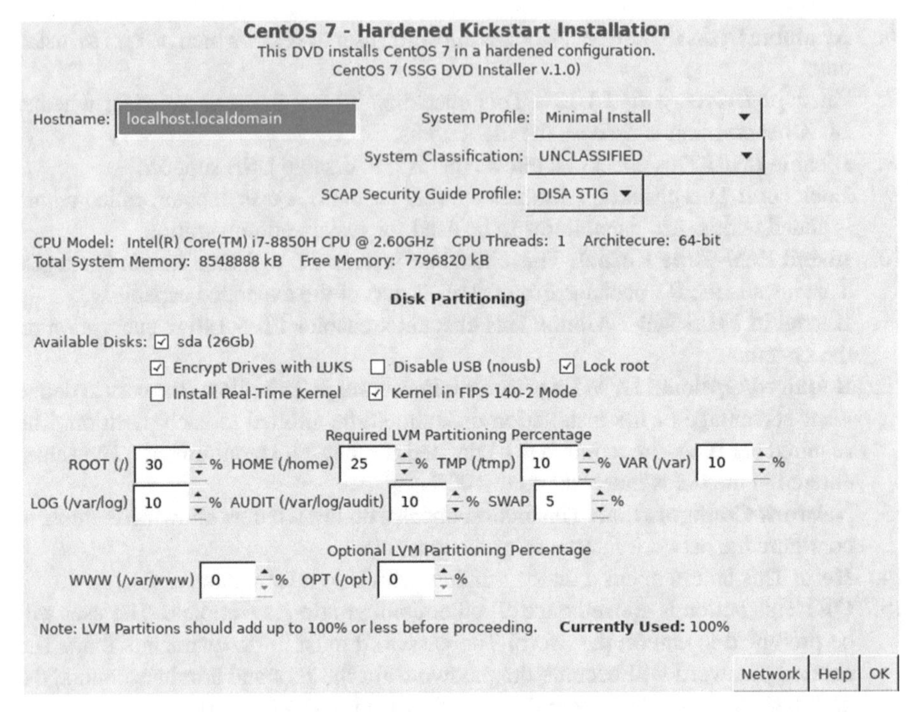

Fig. 2. CentOS hardening installation menu

- **SSSD/realmd:** System Security Service Daemon provided to support centralized authentication mechanisms (i.e. Windows Active Directory).
- **Classification Banner:** A configurable graphical classification banner for Gnome desktops on systems that implement a graphical user interface. The banner occupies a small amount of space at the top and bottom of the screen indicating the classification level of your system. This is an open source project maintained by the Red Hat community and has become common use in Industry [10].
- **USB Guard:** Software to control access to USB devices by implementing basic whitelisting and blacklisting capabilities based on device attributes [11].

4.6 System Profiles

The hardening kickstart file (/kickstart/hardened-centos.cfg) defines multiple profiles (/hardening/profiles/) that specify which packages and settings get installed. This aids in abstracting and specifying those profile configurations for ease of use, viewing, and modifying. These profiles are modifiable, new profiles can be added, and existing profiles removed easily. Except for the Real-time Host profile, all other profiles listed here are from Caviggia's original project. Those additional profiles have not been evaluated for SCAP compliance.

The following profiles are defined within the scope of this system:

- **Minimal Install:** This profile implements a minimum subset of packages during installation. Graphical support (`Xwin`) is not included.
- **Real-time Host:** This profile is designed to be used in a real-time environment as either a development or production host computer. Graphical support (`Xwin`) is included with this profile. The focus of this work and associated analysis occurs within this profile.
- **IPA Authentication Server:** This profile implements the minimal package set plus the packages needed to support IPA Authentication Server (Red Hat Identity Manager).
- **Ovirt KVM Server:** This profile implements the packages necessary to allow Ovirt Manager to be installed on the system. After installation, the script ovirt-engine-install.sh must be run to finish installing the Ovirt Manager packages.
- **Standalone KVM Server:** This profile installs the packages necessary for a standard KVM Virtualization Server.

4.7 Installation and Automation

The final ISO output is installed to the target machine using bootable media in the same form as the standard CentOS ISO distribution and works in the same manner. The ISO will fit on a standard DVD. For larger ISOs that include large amounts of packages and data, a Blu-ray or USB bootable media may be required. When the media is booted the user is presented with the following menu selection.

Selecting Install CentOS 7 Secure Host Baseline will start the standard Linux SHB installation kickstart (/kickstart/hardened-centos.cfg). After making the selection, the user will be presented with the SHB installation menu (see Fig. 2). Once the menu selections are complete, the installation with continue automatically until the system is ready for reboot.

5 Customizing the Distribution

The Linux Secure Host Baseline distribution is designed to be fully customizable and configurable to meet the organization requirements for the deployed system. The two primary methods for customizing the distribution is for changing profiles or kickstart files.

5.1 Profiles

Currently, the primary usage for the individual profiles is to ensure a specific configuration is accomplished during the installation process. The configurations can be modified through the /hardening/profile_package.cfg file and its associated profile Python configuration script (i.e. /hardening/profiles/Realtime_Host.py). The Real-time Host profile configures the disk partitions, runs the hardening and remediation scripts using OSCAP, configures the firewall, and reinstalls `Xwin` for the graphical interface. This method facilitates the final configuration for the system. New profiles can be generated that utilize different configuration settings which can be selected during the installation process to allow for variations in systems.

5.2 Custom Kickstart Files

User defined kickstart files can be added to the kickstart directory to accomplish modifications that cannot be as easily completed within the profile modifications. It is recommended that the user make a copy of and modify the existing /kickstart/hardened-centos.cfg file.

6 Maintaining the Distribution

Security compliance is a constantly evolving process that requires iterative improvements and is rarely static. There are a multitude of events that could trigger improvements or reevaluation over time within the realm of cyber security depending on the deployed environment.

6.1 Trigger Events

The primary trigger events that are specific to this automated Linux SHB approach but does not directly apply to the general state of cyber security are listed below.

- New Release of CentOS.
- New Version of the Red Hat STIG is Released.
- New Version of the SCAP Compliance Checker (SCC) Tool or OpenSCAP (OSCAP) Tool is Released.

6.2 Approach for Updates

Required updates following the trigger events will likely need to occur to centos_rt_shb.py or /kickstart/hardened-centos.cfg and its associated hardening scripts to account for changes to the underlying operating system and to implement new remediations found by the SCC. Figure 3 shows an example process to update the distribution.

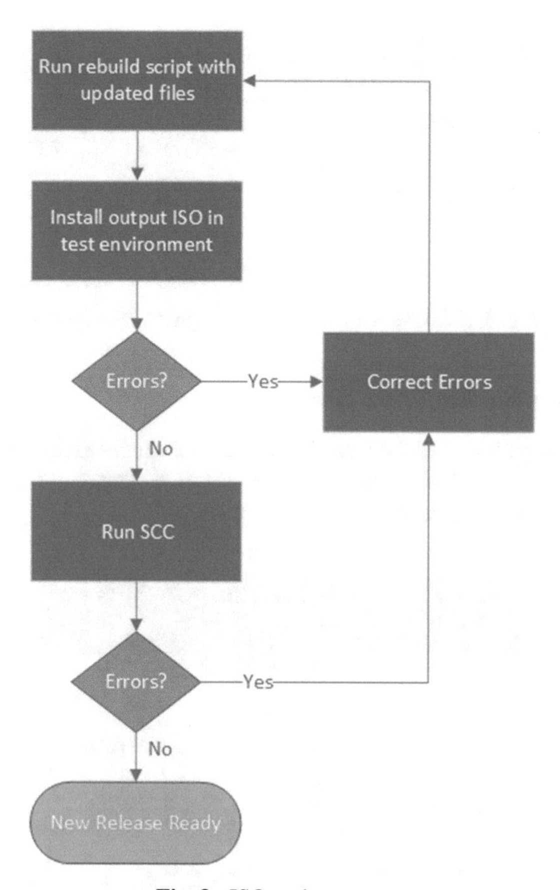

Fig. 3. ISO update process

7 Security Results Analysis

The security compliance of the final system was analyzed using the SCAP Compliance Checker tool and the latest STIG SCAP content at the time that this work was performed, as shown in Fig. 4. The tool uses the SCAP content to evaluate the installed operating system and applications, and generates a report detailing the level of compliance along with deficiencies. The version of the compliance tool used was 5.2 with version 2.3 of the RHEL 7 STIG.

While the use of a security checklist can significantly improve the overall security of the system on which it has been checked against, no checklist can ensure the system is 100% secure. Given that, it is not always possible to reach 100% compliance within the chosen checklist depending on the nuances of a given system. This approach provides an approach to strive toward that 100% compliance, but not necessarily reach the 100% compliance goal.

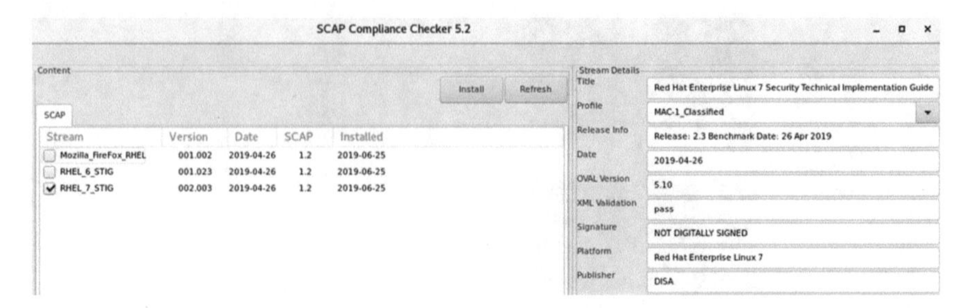

Fig. 4. SCAP Compliance Checker (SCC) information

7.1 Post Remediation Real-Time Host Scan

The post remediation scan resulted in 175 out of 182 checks passed for 96.15% compliance for the Real-time Host profile as shown in Fig. 5. The qualitative guideline puts this into the green compliance status for being above 90%.

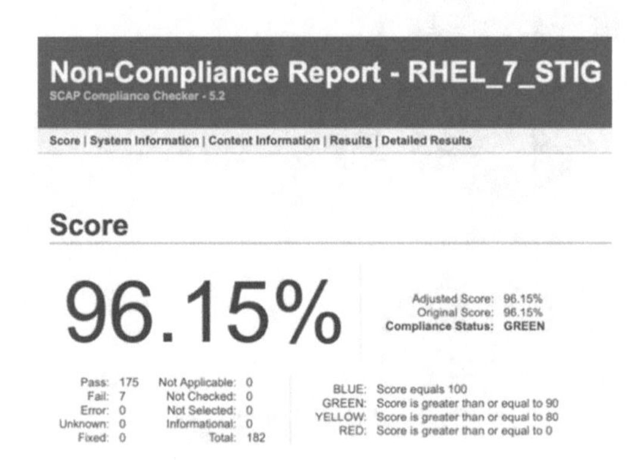

Fig. 5. SCAP compliance metrics

7.2 STIG Deficiencies not Remediated

The STIG items listed in Table 1 were not remediated but their associated STIG ID and description are listed for reference.

UID 1–4 check fail due to the specific operating system used. This approach is using CentOS as opposed to RHEL. The checks are looking for packages that are only provided with REHL and cannot be remediated while using CentOS.

UID 5 fails due to X Windows display manager being installed. The profile that is being used requires X Windows to be installed for graphical user interfaces used within real-time processes. This can be remediated for profiles that do not require X Windows for normal operations.

UID 6–7 failures have unclear causes. The descriptions provided based on the STIG information should allow these checks to pass. So, more investigation would be required to remediate the failures.

Table 1. STIG non-compliance deficiencies

Unique ID (UID)	STIG ID	Description
Non-Red Hat Release		
1	SRG-OS-000480-GPOS-00227	The Red Hat Enterprise Linux operating system must be a vendor supported release. - (CCE-80349-4) – Fail
2	SRG-OS-000033-GPOS-00014	The Red Hat Enterprise Linux operating system must implement NIST FIPS-validated cryptography for the following: to provision digital signatures, to generate cryptographic hashes, and to protect data requiring data-at-rest protections in accordance with applicable federal laws, Executive Orders, directives, policies, regulations, and standards. - (CCE-80359-3) – Fail
3	SRG-OS-000033-GPOS-00014	The Red Hat Enterprise Linux operating system must use a FIPS 140-2 approved cryptographic algorithm for SSH communications. - (CCE-27295-5) – Fail
4	SRG-OS-000250-GPOS-00093	The Red Hat Enterprise Linux operating system must be configured so that the SSH daemon is configured to only use Message Authentication Codes (MACs) employing FIPS 140-2 approved cryptographic hash algorithms. - (CCE-27455-5) – Fail
X Windows Installed		
5	SRG-OS-000480-GPOS-00227	The Red Hat Enterprise Linux operating system must not have an X Windows display manager installed unless approved. - (CCE-27218-7) – Fail
Reason for Failure Unclear		
6	SRG-OS-000327-GPOS-00127	The Red Hat Enterprise Linux operating system must audit all executions of privileged functions. – Fail
7	SRG-OS-000023-GPOS-00006	The Red Hat Enterprise Linux operating system must display the Standard Mandatory DoD Notice and Consent Banner immediately prior to, or as part of, remote access logon prompts. - (CCE-27314-4) – Fail

7.3 Limitations and Proposed Next Steps

While the results of this work highlight the initial capability, the proposed solution is only a snapshot within a single instance in time. As Cyber Security is an ever-changing goal, and the software utilized is being continuously maintained and updated, there will need to be future engineering effort to facilitate improvements over time. For example, CentOS 8 will require some amount of time in deployment to ensure it has been fully vetted, as generally the newest operating systems are not used immediately in production and security environments, which will then require additional modification and testing to use this approach.

The proposed next steps for refinement include the following.

- **System Updating:** Determine the exact CentOS, RHEL STIG, and SCAP Compliance Tool updating impacts and their associated update process for affected files and scripts.
- **Patch Updating:** Determine the exact update process for incorporating the latest patches rather than generating a completely new ISO image.

- **Red Hat Enterprise Linux 7 Support:** Support for RHEL 7 should be included in the scripts. Some organizations require or will choose RHEL over CentOS, and this should only require minimal modification.
- **Deployment Readiness:** Additional testing and refinement to develop the scripts, kickstart files, and their support files should be performed to ensure the system is ready for deployment.
- **Feedback:** Iterative update processes from security administration professionals would be invaluable, as well as other use cases to generate additional profiles.
- **Additional Security Applications:** Determine the impact from including security applications that improve security posture such as a Domain Controller, Virus Scanner (i.e. McAfee), and System Logger (i.e. Kiwi Syslog).

8 Conclusion

The utilization of Linux is highly preferred in systems requiring real-time performance constraints. As generally real-time systems are used within industrial, aviation, and safety-critical applications, there is an inherent need or requirement to achieve a certain level of security compliance. This process can often be manual and require expertise to achieve. A goal of this work is to provide a framework to improve that current process, apply it using cost effective tools, automate the creation and distribution specific to real-time applications, and to quantitatively measure the resulting security posture of the deployed system.

The results of the SCAP Compliance Checker indicate a high level of compliance for the automated CentOS Real-time Host profile using the DISA RHEL 7 STIG. The approach presented offers an automated Linux solution to match the intent for improved security as proposed by the DoD approved Windows Secure Host Baseline while offering real-time functionality through the inclusion of the PREEMPT_RT patch.

References

1. R. O. Work: Implementation of Microsoft Windows 10 Secure Host Baseline. Official memorandum, Department of Defense, Washington, D.C., USA (2016). https://dodcio.defense.gov/Portals/0/Documents/Cyber/DSD%20Memo%20-%20Implementation%20of%20Microsoft%20Windows%2010%20Secure%20Host%20Baseline.pdf
2. Quinn, S., Souppaya, M., Cook, M., Scarfone, K.: National Checklist Program for IT Products - Guidelines for Checklist Users and Developers. NIST Special Publication (SP) 800-70 Revision 4 (2018)
3. NIST Joint Task Force: Security and Privacy Controls for Federal Information Systems and Organizations. NIST Special Publication 800-53 Revision 4 (2013)
4. Arch Linux: Realtime kernel patchset. https://wiki.archlinux.org/index.php/Realtime_kernel_patchset
5. Linux Foundation: Real-Time Linux. https://wiki.linuxfoundation.org/realtime/start
6. Center for Development of Security Excellence: Getting Started with the SCAP Compliance Checker and STIG Viewer Job Aid. https://www.cdse.edu/documents/cdse/SCAP-compliance-checker-and-STIG-viewer-job-aid.pdf

7. Space and Naval Warfare (SPAWAR) Systems Center Atlantic (SCC-LANT): SCAP Compliance Checker User Manual for Red Hat Enterprise Linux, Version 4.0 (2015)
8. Caviggia, F.: Hardened CentOS 7 Kickstart, January 2019. https://github.com/fcaviggia/hardened-centos7-kickstart
9. Red Hat: Hardened RHEL 7 Kickstart, August 2017. https://github.com/RedHatGov/ssg-el7-kickstart
10. Cavigga, F.: Classification Banner, September 2018. https://github.com/fcaviggia/classification-banner
11. Red Hat: USBGuard (2019). https://github.com/USBGuard/usbguard
12. Henttunen, K.: Automated hardening and testing CentOS Linux 7: security profiling with the USGCB baseline (2018)

Deficiencies of Compliancy for Data and Storage

Isolating the CIA Triad Components to Identify Gaps to Security

Howard B. Goodman[1(✉)] and Pam Rowland[2]

[1] Dakota State University, 820 Washington Avenue N, Madison, SD 57042, USA
howard.goodman@trojans.dsu.edu

[2] Dakota State University, SD 820 Washington Avenue N, Madison, SD 57042, USA
pam.rowland@dsu.edu

Abstract. There are many reasons to implement data protection security strategies regardless of if the data is financial, personal or confidential, risks are a moving target. The purpose of this systematic literary review was to examine articles and documents pertaining to data and storage security and to compare with modern regulatory compliance requirements to determine if gaps exist within the datacenter. Both academic and applied IT security papers were used as well as online governmental and industry sources. As part of this research, the components of the CIA triad were used as a baseline which resulted in a granular model. The model was applied to both quantitative and qualitative data that exposed deficiencies in data and storage security.

Keywords: Storage · Data · Protection · Security · Compliance · CIA · Disaster recovery · Business continuity · Access controls · Personal information · Breach

1 Introduction

At the core of every organization is the datacenter and for the most part, the technology that is relied on to maintain operations. Information Technology is shifting from being a tool that helps the operations to a gateway of the most precious commodity an organization owns, data.

Data and storage can no longer be siloed from one another as the security of data is dependent on its location, form, and how it's accessed. The objective of this paper is to conduct a systematic literary review of data and storage security from both academic and industry sources, correlated to "Confidentiality, Integrity and Availability" ("CIA" or "CIA triad") and determine if gaps in data and storage security exist.

1.1 Prevailing Definitions

Over the years the term "datacenter" has varied slightly, but for the purpose of this paper we will think of a datacenter as a physical or virtual space where information systems and data reside for an organization [1]. They can be a local or remote physical shared space or even a managed service and the level of responsibility to support and maintain the

K.-K. R. Choo et al. (Eds.): NCS 2020, AISC 1271, pp. 170–192, 2021.
https://doi.org/10.1007/978-3-030-58703-1_11

datacenter can vary widely from organization to organization [2]. With so many concerns for critical backend services, it is logical to consolidate the security, environment, power and other requirements to a central location. Regardless of location or physicality, the datacenter is where applications and data are accessed [2].

Data has many forms, but for the purposes of this paper, we will classify data as either:

- Traditional structured or unstructured data
  - Structured data is data that is stored in a predefined format and organized so that it can be referenced; such as a database [3]
  - Unstructured data is data in the form of flat files that are not in a predefined format such as a text file, pdf, image or typically some file that would be accessed by users directly [4].
- Created data - gathered or the result of an organization's information systems, applications or users and stored i in the form of either structured or unstructured data
- Virtual Machine (VM) data – this would include files for the metadata that defines VM resources - virtual disks stored as a file that hold the OS, installed applications and can have structured and unstructured data.

Data must be stored and secured somewhere and available for use. The emphasis of data storage security utilized for backend operations is identified as utility-based primary storage that is directly used for computer services. This type of storage can be in any of these forms:

- Storage Area Networks (SAN)
- Network Attached Storage (NAS)
- Direct Attached Storage (DAS)
- Software Defined Storage (SDS)
- Object Oriented Storage (OOS) or Object-based Storage Devices (OSD)
- Content Addressable Storage (CAS)
- Cloud Storage (OneDrive, Dropbox, Google Drive, Box, etc.).

Virtualization has paved the way for the next wave of the datacenter. In the age of the Cloud's "As-A-Service", datacenter storage has emerged into three primary classifications which embody functional requirements. [5] These primary functional types are:

1. **Block storage** – storage presented to systems as raw disk that can be controlled and formatted by host [6].
2. **File storage** - hierarchical storage where files are organized under directories and are presented to host as SMB/CIFS or NFS protocol [7]
3. **Object storage** – storage that is separated into 3 parts: [8]

a. The data (can be almost anything)
b. Expandable metadata – who/when created and any other relevant information
c. Global unique identifier.

Also, archive storage has been identified as a type of storage, but it is typically some type of nearline storage used for a specific use-case, such as stagnant offline or long-term storage and can be a subset of one or more of the 3 primary function types listed above. In addition, there is data transport storage, but this is related to inter-communication of storage or data. The 3 classifications represent most of the use cases needed for enterprise datacenter requirements. The simplification defines the nomenclature based on functionality without the concern for the underlying technology or hardware. Regardless of the storage technology or functional type, there is a need for standardization of data and storage security.

1.2 Compliancy Standards

Counterintuitive to the definition of "standards", compliance can vary based on industry and type of data. The following list is a quick overview of the most common primary security compliancy frameworks with its purpose and related data and storage security requirements.

- **European Union's General Data Protection Regulation (GDPR)**
  - **Purpose:**
    - Protect European's personal information used or stored by any business worldwide [9]
  - **Requirements:**
    - Data needs to be encrypted [10] and protected with penalties if misused or neglected [11]
    - Enforcement of any organization with Personally Identifiable Information (PII)
    - Uses an approved method to secure data [11]
    - A "right to know law" – companies are required to inform appropriate organizations when consumer data is breached [11]
- **Payment Card Industry Data Security Standards (PCI DSS)** [12]
  - **Purpose:**
    - Protect Cardholder Data (Visa, MC, AMEX, Discover) and PII
  - **Requirements:**

- Defined how PCI data should be secured in transit and how it's secured when stored
- Requirements for security and updates of OS and security applications
- How it is protected (i.e. backed up and BC/DR is implemented) [10]
- Implementation of strong access controls
- Must also have an incident response plan [13]

- **U.S. Health Insurance Portability and Accountability Act of 1996 (HIPAA)**
  - **Purpose:**
    - To protect patient's use and disclosure of health records and patient information [14]
  - **Requirements:**
    - Focus on privacy security and access controls of medical records [14]
    - Continuous availability of data [14]
    - Notifications of breaches and data that is lost or stolen [14]
- **Sarbanes-Oxley Act of 2002 (SOX)**
  - **Purpose:**
    - Law created to cut down on corporate fraud [15]
  - **Requirements:**
    - It bans companies from giving loans to executives, protects whistleblowers and requires the CEO certify the accuracy of financial information and is punishable for up to 20 years for false filing [15]
    - Requires periodic auditing [15]
    - Requires data retention for at least 7 years [10]
- **Federal Risk and Authorization Management Program (FedRAMP)**
  - **Purpose:**
    - Create guidelines for federal agencies use of cloud computing [10]
  - **Requirements:**
    - Standards for security assessments, authorization and monitoring cloud offerings [16]
    - Required for all federal agencies using cloud services [16].

In addition to these compliance regulations, many individual countries have their own laws and every state in the U.S. has some type of data breach or data protection law to protect personal information increasing the complexity of data security and the concern of the public. [17–19] For example, one of the most recent being the California Consumer Privacy Act (CCPA), which was introduced in January 2018 and came into effect in January 2020. [20] CCPA is notable because it is most closely correlated to the EU's GDPR with some exceptions based on scope and reach [20].

By examining the requirements for the different compliances, we can see some overlap, but each has some variants and commonalities with the NIST cybersecurity framework. According to NIST, organization's security emphasis is most concerned with the Confidentiality aspect of CIA and often overlooks the other aspects [21].

2 Research Methodology

In order to complete this systematic review of storage and data security, it seemed logical to keep the regulatory requirements and the key concepts of the CIA tirade in mind throughout this review. As a result, we were able to target keywords, themes and concepts for data and storage to measure what areas of CIA are addressed and to identify what is missing from the research.

2.1 Sources of Articles

Searches were performed from September 2019 to November 2019 and 100 articles were reviewed with 81 documents selected (see references for detail list). Documents and articles were found using the Dakota State University's Karl Mundt Library's research databases which gave access to the following resources that were used throughout this study:

- ACM Digital Library
- IEEE Xplore Digital Library
- American Council for an Energy-Efficient Economy, or ACEEE
- Google Scholar
- InfoSecurityNetBase
- ProQuest Research Library
- National Technical Information Service (NTIS).

In addition, well known standards organizations were also researched. Notable and relevant sources included:

- Storage Networking Industry Association (SNIA) – Publicly available
- National Institute of Standards and Technologies (NIST) – Publicly available
- International Organization for Standardization (ISO) – Available through ANSI University Outreach Program
- Payment Card Industry Security Standards (PCI SSC) – Publicly available
- U.S. Department of Health and Human Services (HHS) – Publicly available.

Articles were eliminated for review:

- when they showed bias
- when they were used as an advertisement or self-promotion
- if they were written before 2005.

Academic sources needed to be published and available on one of the library's research databases and industry sources needed to be published by an industry professional or organization.

2.2 Security Principles

During this literary review, the key terms and ideas of the research were tracked and as a result, seven key security terms emerged. Below we identified these security principles that are fundamental to data and storage security:

1. **Authentication:** Access control for validating that access to data and storage is allowed
2. **Authorization:** Access control for management and governance of authentication
3. **Privacy:** Ensure data and storage is isolated, encrypted and allowed or decrypted by valid source of authority
4. **Reliability:** Ensure data is accurate and storage is durable and working as designed
5. **Verification:** Auditing, inspection and analysis of data and storage
6. **Recoverability:** Ability for data and storage to return to a good or known state
7. **Accessibility:** Data and storage can be reached and usable as intended.

These seven security principles can be correlated to CIA as followed (Table 1):

Table 1. 7 Principles cataloged under CIA triad

Confidentiality	Integrity	Availability
• Authentication • Authorization • Privacy	• Reliability • Verification	• Recoverability • Accessibility

2.3 Source Classification

Using this as a baseline for cataloging articles, a search for both academic and industry was conducted from various sources for both data and storage security standards, best practices and recommendations. The method was developed over the course of 3 months and was tracked using Excel where the articles were sorted by the seven defined security principles.

This became a working spreadsheet that gave way to a table that traced the driving themes of the research reviewed. While many of the sources fell into multiple categories, the main or principle theme was the single classification or primary theme of the article that was tallied.

3 Article Mapping to CIA Triad

By using the chart from Table 2, the findings were organized based on the primary category of the document. This security taxonomy allowed summation of all the reviewed documents which formed an interpretation that produced observable patterns. For readability, the CIA breakdown is shown in the below 3 tables. (See Table 2, Table 3 and Table 4),

Table 2. Confidentiality

Confidentiality			
Totals:	30		
Security principle	Authentication	Authorization	Privacy
Storage security	4	6	4
Data security	1	4	11
Academic	3	3	5
Industry	2	7	10
(First Author, year) [reference #]	1. (SNIA 2018) [22] 2. (Schopmeyer 2017) [23] 3. (Smith 2016) [24] 4. (Daniel 2019) [25] 5. (Park 2015) [26]	1. (Butler 2008) [27] 2. (Tang 2018) [28] 3. (Hibbard 2016) [29] 4. (Willett 2016) [30] 5. (SNIA 2015) [31] 6. (SNIA 2016) [32] 7. (SNIA 2015) [33] 8. (McKay 2019) [34] 9. (ENISA 2019) [35] 10. (Zhou 2019) [36]	1. (PCI SSC 2015) [12] 2. (PCI SSC 2017) [9] 3. (Sarkar 2014) [37] 4. (Krahn 2018) [38] 5. (Hibbard 2014) [39] 6. (Hibbard 2014) [40] 7. (SNIA 2014) [41] 8. (SNIA 2018) [42] 9. (PCI SSC 2018) [43] 10. (PCI SSC 2010) [44] 11. (Schaffer 2019) [45] 12. (Brandão 2019) [46] 13. (Zyskind 2015) [47] 14. (Wang 2019) [48] 15. (Jie 2014) [49]

Table 3. Integrity

Integrity		
Totals:	27	
Security principle	Reliability	Verification
Storage security	13	6
Data security	4	4
Academic	6	7
Industry	11	3
(First Author, year) [reference #]	1. (Jovanovic 2010) [50] 2. (Butler 2007) [51] 3. (Xu 2005) [51] 4. (Paik 2018) [52] 5. (Hibbard 2015) [53] 6. (Arnold 2011) [54] 7. (ISO 2015) [55] 8. (SNIA 2017) [56] 9. (SNIA 2016) [57] 10. (SNIA 2015) [58] 11. (SNIA 2012) [59] 12. (SNIA 2009) [60] 13. (SNIA 2010) [61] 14. (Gordon 2019) [62] 15. (Talib 2010) [63] 16. (Dharma 2013) [64] 17. (IBM 2019) [65]	1. (Hasan 2006) [66] 2. (Vasilopoulos 2018) [67] 3. (Zhu 2010) [68] 4. (PCI SSC 2018) [13] 5. (Sarkar 2017) [69] 6. (Hou 2019) [70] 7. (Schulz 2011) [71] 8. (Kwon 2018) [72] 9. (DellEMC 2018) [73] 10. (HDS 2019) [74]

For readability, only the first author and year published is recorded in the three tables, however, the reference numbers are listed in order and all source documents are cited.

4 Results of Classification

The classification for each article and document was done independently to avoid bias to shape the results. The breakdown based on the CIA Triad was relatively evenly distributed with 30% Availability, 33% Integrity and 37% Confidentiality. It is also notable that neither the academic nor industry articles focus on one specific security area and were split evenly over data and storage security attentions (See Table 5).

As the research progressed, 6 sources emerged as primary sources. Table 6 shows the top 6 sources that published articles on data and/or storage security:

In addition, the publications can also be seen with the number of articles selected by year (See Table 7):

Table 4. Availability

Availability		
Totals:	24	
Security principle	Recoverability	Accessibility
Storage security	7	5
Data security	7	5
Academic	5	6
Industry	9	4
(First Author, year) [reference #]	1. (Li 2016) [75] 2. (SNIA 2019) [76] 3. (SNIA 2016) [77] 4. (SNIA 2010) [78] 5. (McMinn 2009) [79] 6. (McMinn 2009) [80] 7. (McMinn 2009) [81] 8. (Dutch 2010) [82] 9. (SNIA 2017) [83] 10. (Schopmeyer 2009) [84] 11. (Chang 2009) [85] 12. (Jian-hua 2011) [86] 13. (Wang 2018) [87] 14. (Bollinger 2015) [88]	1. (Zhou 2014) [89] 2. (Chen 2019) [90] 3. (Carlson 2017) [91] 4. (SNIA 2008) [92] 5. (Fuxi 2015) [93] 6. (Rouse 2019) [94] 7. (BlockApps 2017) [95] 8. (Xu 2018) [96] 9. (Zheng 2019) [97] 10. (Veleva 2019) [98]

Table 5. Source type by concertation of security interest

	Storage security	Data security
Industry	54.35%	45.65%
Academic	57.14%	42.86%

Table 6. Top article sources

Top sources for articles and papers						
SNIA	ACM	IEEE	EBSCOhost	PCI SSC	InfoSecurityNetBase	Other
29	12	8	7	5	4	16

Table 7. Year published grouped over 3-year increments

Years published				
2005–2007	2008–2010	2011–2013	2014–2016	2017–2019
3	15	5	22	36

From here we can observe the following notable results:

- 35 of the articles were published through academic sources (43%)
- 43 articles were from industry sources (57%)
- Over 60% of the documents were produced after 2015
- The leading source (>35%) of publications was from SNIA.

We can see that comparison of academic and industry articles cataloged by the 7 defined security principles. (See Fig. 1)

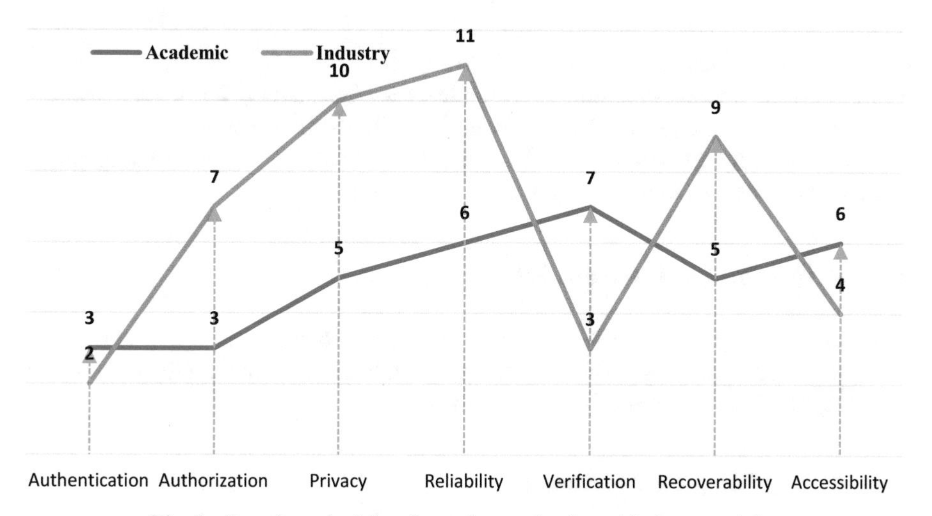

Fig. 1. Security principles shown by academic and industry articles

What we notice is the following:

- Industry focuses more on reliability (11), privacy (10) and recoverability (9)
- Academic focuses mostly on verification (7), reliability (6) and accessibility (6)
- The least focus for both was around authentication.

In Fig. 2 we can see the comparison based on storage and data security.
Here we can make the following observations:

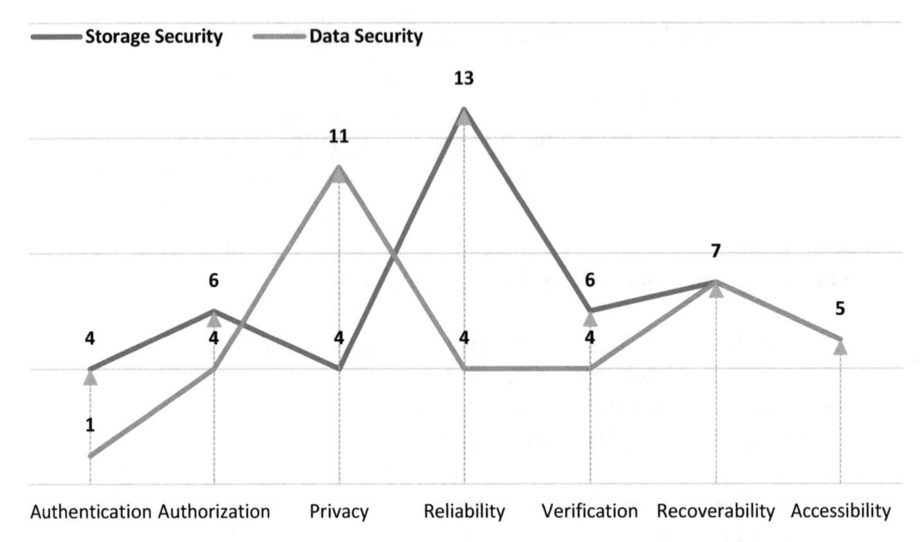

Fig. 2. Security principles compared by data and storage articles

- Data security articles focus mostly on privacy (11)
- Data security articles were limited on the Integrity (Reliability, Verification) aspect
- Storage, on the other hand, focused on Integrity (Reliability, Verification)
- Storage security articles only touched on confidentiality with most of the emphasis on access controls (Authentication and Authorization).

5 Analysis and Interpretation

By utilizing this finite view, we can apply these terms to gain a more granular understanding of where the gaps to data and storage exist. In this section we will examine how we can apply these 7 classifications to real-world data.

5.1 Industry Perspective of Storage

When we look at modern storage today, we see common security solutions that have many of the same characteristics, capabilities and features. This in turn has enabled technical advances for organizations of all sizes, budget, expertise or location. In addition, the technical advances made with capacity, performance and resilience have made it possible to see massive consolidation. Even the largest enterprise organizations will have datacenters that are 75% or more virtualized [99]. For this reason, looking at virtualization gives us significant insight that touches most organizations today.

In a 2015 survey conducted by DataCore Software, IT professionals were asked: "What are the primary reasons your organization chose to deploy storage virtualization software?" The 477 respondents highlighted the need for flexibility, business continuity, disaster recovery or cost reduction [100] (See Fig. 3).

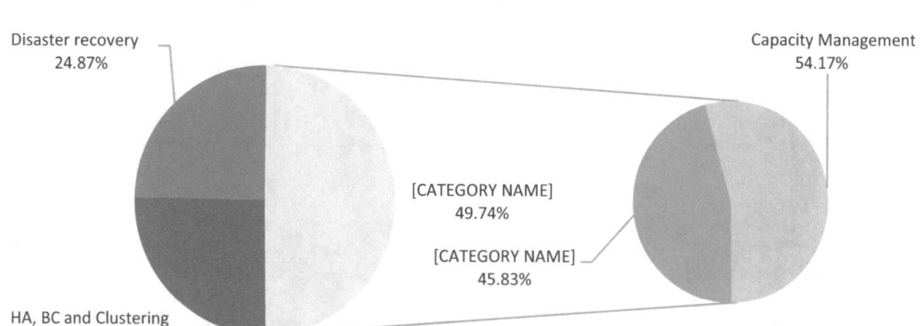

Fig. 3. Primary reasons for deploying storage virtualization software 2015 [100]

What this shows is that back in 2015, the security concern that virtualization was valued for, was:

- Recoverability
  - 24.87% - "Disaster recovery"
- Accessibility
  - 25.39% - "High availability, business continuity, metro clustering".

This shows that at that time, the principles based on confidentiality and integrity were not deciding factors for virtualization.

In another study conducted from 2016 to 2017 by 451 Research, IT decision-makers were asked what the top concerns were regarding data storage. The top 4 concerns were: Capacity, Management, Cost and Performance [101].

In the below figure (see Fig. 4), we see an abbreviated version of 13 areas that were considered the leading causes of problems for data storage from 2016–2017 [101].

If we map this to the 7 security principles, we start to see more portions of CIA:

- Authorization
  - Management – 2016 (31%) and 2017 (50%)
- Verification
  - Compliance – 2016 (14%) and 2017 (16%)
- Recoverability
  - Disaster Recovery – 2016 (28%) and 2017 (31%).

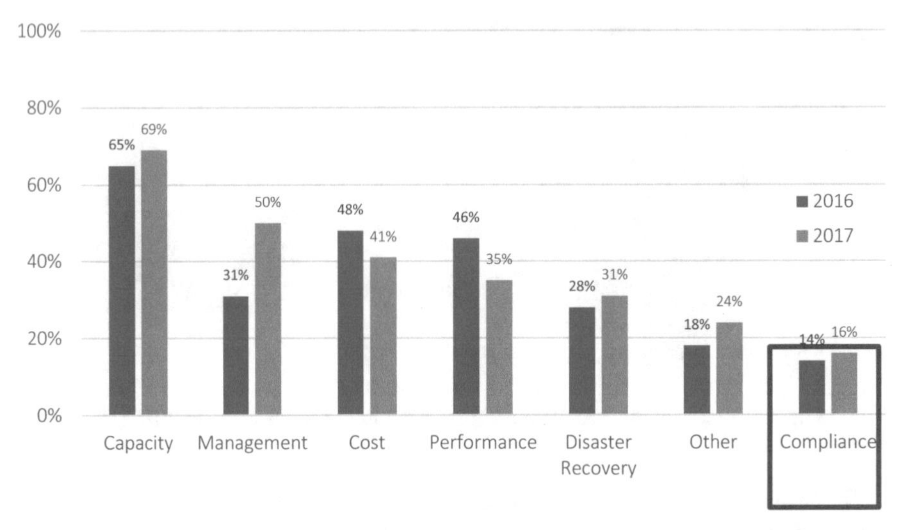

Fig. 4. Leading problems for data storage worldwide in 2016 and 2017 [101]

This is a step in the right direction, but still major gaps showing functional requirements are far more of a value then security concerns and greater then 80% don't consider compliance a problem related to data storage. [101]

5.2 Data Risks

In the 2018 European Union Agency for Cybersecurity (ENISA) Threat Landscape Report 2018, we can see that IT assets vulnerable to insider threats (see Fig. 5), show that the top targets are the containers for structured and unstructured data. [102]

The same ENISA report, shows that the data types vulnerable to insider threats (See Fig. 6) fall into the categories of data that is regulated. [102]

In Fig. 6, the top 3 data types at risk are:

1. Confidential business information
2. User access credentials
3. Protected Health Information (PHI) and Personally Identifiable Information (PII)

From both of these studies we are starting to see a greater appreciation for applying the 7 principles as we are more than ever concerned with device level access controls, system and application controls as well as the different forms of data confidentiality and integrity.

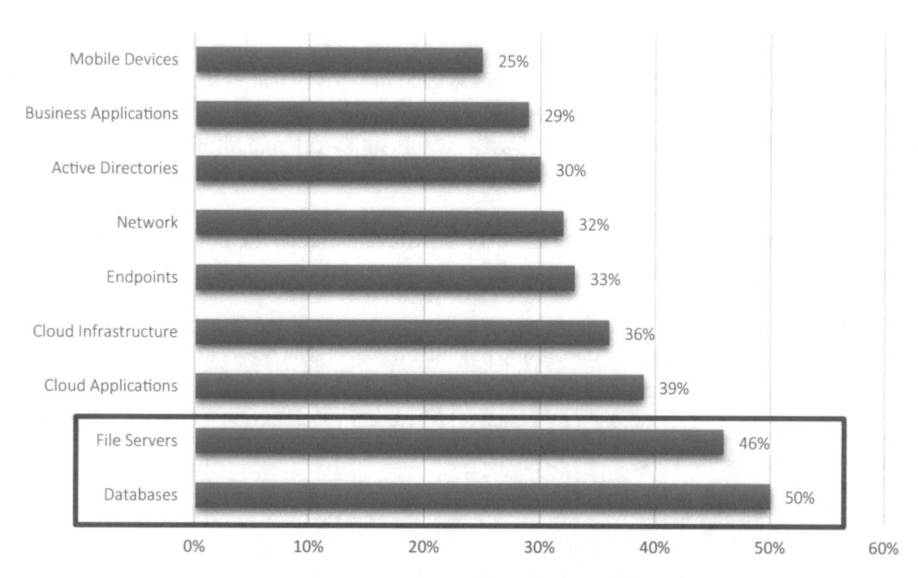

Fig. 5. IT assets vulnerable to insider threats [102]

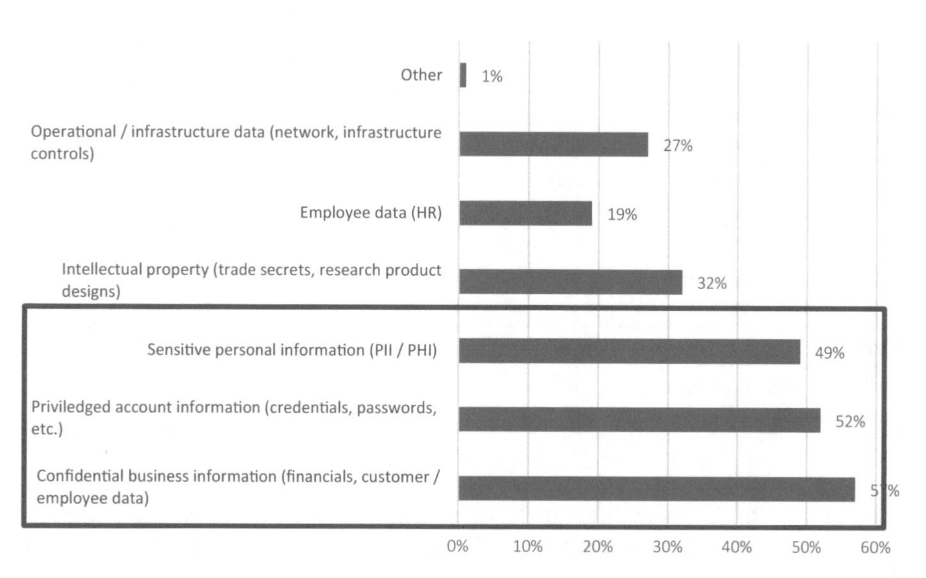

Fig. 6. Data types vulnerable to insider threats [102]

5.3 Security Threats Related to Data Storage

In order to fully understand the threats that can be the target for data storage we need to examine the outcome we are trying to avoid. The threats can be active, passive or incident based. Active means a threat-agent is actively trying to create a result such as denial of service, repudiation or modification attack. Passive indicates that the threat-agent is trying to gain access or read or analyze stolen data or transmission [103]. Incident based threats are those that require partial or full recovery of one or more critical systems. The end result will fall into one or more of 6 areas where unlawful, unauthorized or accidental incidents can occur against or effecting data disclosure, alteration, destruction, loss, corruption and/or access: [55]

The below table maps the data security threats and shows the potential form of the data breach. (See Table 8).

Table 8. Security threats and potential forms of data breaches [55]

	Access	Disclosure	Alteration	Destruction	Loss	Corruption
Theft of storage element or media	X	X	–	X	X	–
Loss of storage element or media	X	X	–	X	X	–
Loss of data	–	–	–	X	–	X
Accidental configuration changes by authorized personnel	X	X	X	X	–	–
Malicious configuration changes by external or internal or adversaries	X	X	X	X	–	–
Privileged user abuses by authorized users	X	X	–	–	–	–
Malicious data tampering by external or internal adversaries	–	–	X	X	–	–
Denial of service attacks	–	–	X	–	X	–
Malicious monitoring of network traffic	–	X	–	–	–	–

We can then take the different forms of breaches and use the 7 principles to identify which are needed as counter measures. In Table 9 we see which security principles need to be applied.

Table 9. Potential forms of data breach mapped to 7 security principles

	Confidentiality			Integrity		Availability	
Potential forms of data breach	Auth.	Author.	Priv.	Reliab.	Verif.	Recover.	Access.
Unlawful, unauthorized or accidental access	X	X	X	–	X	–	–
Unlawful, unauthorized or accidental data disclosure	X	X	X	–	X	–	–
Unlawful, unauthorized or accidental data alteration	X	–	–	X	X	–	–
Unlawful, unauthorized or accidental data destruction	–	–	–	–	X	X	–
Unlawful, unauthorized or accidental data loss	–	–	–	–	X	X	X
Unlawful or accidental data corruption	–	–	–	X	X	–	X

5.4 Regulatory Requirements for Disclosure of Data Breaches

We are limited to known data losses or incidents that are either publicly announced, because they affect too many individuals to keep secret, or are required to be reported for some compliance reason. There is also the possibility that an organization may be unlikely to voluntarily report incidents to avoid the social stigma or embarrassment. Lastly, storage and data protection companies that may have direct sight into incidents would not volunteer this data for a number of reasons such as loss of customer confidence or a mandate of a Nondisclosure Agreement (NDA).

However, since HIPAA 1996 includes a breach notification rule which requires that relevant healthcare organizations notify individuals and the U.S. Department of Health & Human Services (HHS) of violations to PHI we can include the data for analysis [104]. In the below chart we can see a summary view of the types of breaches organized by the year reported, number incidents of the type and the number of individuals affected (See Table 10).

In addition to the breach types, we observe that the HHS required that the reporting organization identify the location. The below chart shows the breakdown for each of the years recorded. It should be noted that a breach could have 2 or more locations, but they didn't show any further division. For example, if a breach occurred with Email and Network Server, it didn't state what portion was Email (See Table 12).

These results are quite alarming as they show that over the past 3 years the amount of data loss incidents being reported is growing exponentially as healthcare providers adhere to HIPAA's notification rule. This is not to say that before 2019 there were less breaches, but rather a lack of reporting or known incidents. Looking at each of these breach types, we compare each breach to the security principles and determine an underlying view based on what we know of the nature of the breach (See Table 11).

Table 10. Reported breaches from since 2017 HHS: # individuals and # incidents [105]

	2017		2018		2019	
Breach Types	Individuals affected	Incidents	Individuals affected	Incidents	Individuals affected	Incidents
Hacking/IT incident	1,054	2	4,470,460	89	35,450,582	256
Unauthorized access/disclosure	31,422	2	2,882,560	44	4,555,857	115
Theft	0	0	70,113	18	354,016	31
Loss	0	0	2,917	3	71,067	12
Improper disposal	0	0	318,811	5	21,907	4
Totals:	**32,476**	**4**	**7,744,861**	**159**	**40,453,429**	**418**

Table 11. Rating the breach types to the 7-security principle

Breach types	Authentication	Authorization	Privacy	Accessibility
Hacking/IT incident	Failed	Failed	Failed	Failed
Unauth. access/disclosure	Failed	Failed	Failed	Failed
Theft	Failed	Failed	Failed	Failed
Loss	Failed	Failed	Failed	Failed
Improper disposal	N/A	N/A	Failed	N/A

Table 12. The data loss reported included at least one or more percent of the total incidents

	2017	2018	2019
Location of breached information	# Incidents	# Incidents	# Incidents
Email	75%	36%	43%
Electronic medical record (EMR)	25%	8%	7%
Other	0%	13%	14%
Paper/films	0%	14%	11%
Network server	0%	25%	26%
User endpoint devices	0%	26%	15%

What we observe here is that most of the types of breaches can be seen as some form of failure with one or more security controls. We also observe that no correlation can be made around reliability or recoverability as we have no way to conclude that any of these principles were relevant. As for verification, you could argue that because the breaches were reported, verification didn't fail, however, the data doesn't tell us when the breach occurred, was discovered and when if any, remediation occurred.

6 Summary

The evolution of data storage has exposed datacenter resources to new threats that require a new model for data protection. This is not to say that the CIA triad isn't relevant, but rather just lacks the granularity that is necessary to identify security needs. The pressure to modernize the datacenter forces IT professionals to choose data aggregated storage to meet functional requirements without taking into consideration the implications to security because guidance and recommendation models are still inadequate. As the virtualized, IP-based and cloud storage technologies continue to become the norm, traditional cyber-attacks are no longer just a networking problem with both outside and inside security threats. And we cannot ignore delineation between datacenter and a front-end network perimeter as this further adds to the security trials.

To add to the problem, wrongfully manipulated data may never be discovered. In the age of Big Data, organizations have influenced trends and behaviors using statistical controls [106]. This indicates a need for greater reliability and validation measures. Beyond these security challenges, compliance regulations continue to expand and have become more complex. What we've seen is that despite compliance, organizations are exposed to security violation. This has been shown with the HIPAA data and with GDPR and other new personal data regulations requiring public notifications, and it's likely that we will see to just what extent data storage is vulnerable. The "shining light" to this problem is organizations and IT professionals are becoming more aware and are seriously looking at solutions to implement.

By applying the seven principles to data and storage we can determine if deficiencies exist within an organization's current IT security policies. As we see in Table 9, we are able to classify the breach types using the seven principles to gain an enhanced view within the CIA Triad. This in turn, can help us focus on only the areas of concern for mitigation. As a result, we can use for educating IT staff and leadership for better planning, management and spending decisions. The more we are able to add clarity the more capable we are to handle future incidents.

In conclusion, future research will include an in-depth study and comparison of each of the regulatory compliancy models as they correlate to the seven principles to determine if deficiencies are quantifiable to support data security. Lastly, with new notification requirements becoming public, we will continue to expand this research and measure the gaps and determine if public notification is a benefit to promote accountability and reduce the unavoidable threats that will continue to plague the risks to individual, personal and customer data.

References

1. What is a Datacenter? Cisco. https://www.cisco.com/c/en/us/solutions/data-center-virtualization/what-is-a-data-center.html. Accessed 13 Dec 2019

2. What is a Datacenter? Definition from Techopedia. Techopedia.com. https://www.techopedia.com/definition/349/data-center. Accessed 13 Dec 2019
3. Beal, V.: What is structured data? Webopedia definition. https://www.webopedia.com/TERM/S/structured_data.html. Accessed 13 Dec 2019
4. Unstructured Data: Wikipedia, 03 December 2019
5. Weins, K.: Compare top public cloud providers: AWS vs Azure vs Google. Flexera Blog, 17 January 2018. https://www.flexera.com/blog/cloud/2018/01/compare-top-public-cloud-providers-aws-vs-azure-vs-google/. Accessed 13 Dec 2019
6. Poojary, P.: Understanding object storage and block storage use cases|cloud academy blog. Cloud Academy, 12 March 2019. https://cloudacademy.com/blog/object-storage-block-storage/. Accessed 13 Dec 2019
7. IBM: File-storage, 14 October 2019. https://www.ibm.com/cloud/learn/file-storage. Accessed 13 Dec 2019
8. Porter, Y., Piscopo, T., Marke, D.: Object storage versus block storage: understanding the technology differences. Druva, 14 August 2014. https://www.druva.com/blog/object-storage-versus-block-storage-understanding-technology-differences/. Accessed 13 Dec 2019
9. PCI SSC: PCI data security standard (PCI DSS). PCI SSC (May 2017)
10. Patterson, C.: Why your current disaster recovery strategy may not cover compliance. Navisite (November 2018)
11. Palmer, D.: What is GDPR? Everything you need to know about the new general data protection regulations. ZDNet. https://www.zdnet.com/article/gdpr-an-executive-guide-to-what-you-need-to-know/. Accessed 16 Dec 2019
12. PCI SSC, "Payment Card Industry (PCI) Data Security Standard." PCI SSC, Jun-2015
13. PCI SSC: The prioritized approach to pursue PCI DSS compliance. PCI SSC (June 2018)
14. Sivilli, F.: What is HIPAA compliance? | Requirements to be HIPAA compliant. Compliancy Group. https://compliancy-group.com/what-is-hipaa-compliance/. Accessed 16 Dec 2019
15. Amadeo, K.: 4 ways sarbanes-oxley stops corporate fraud. The Balance, October 2019. https://www.thebalance.com/sarbanes-oxley-act-of-2002-3306254. Accessed 16 Dec 2019
16. Frequently Asked Questions | FedRAMP.gov. https://fedramp.gov/faqs/. Accessed 17 Dec 2019
17. What's Data Privacy Law in Your Country?: PrivacyPolicy.org, September 2019. https://www.privacypolicies.com/blog/privacy-law-by-country/. Accessed 29 Dec 2019
18. Mulligan, S.P., Freeman, W.C., Linebaugh, C.D.: Data protection law: an overview. Congressional Research Service (March 2019)
19. State Data Breach Law Summary: Baker & Hostetler LLP (July 2018)
20. California Consumer Privacy Act: Wikipedia, 18 December 2019
21. Sebayan, D.: How NIST can protect the CIA triad, including the often overlooked 'I' – integrity. IT Governance USA Blog, Apt 2018. https://www.itgovernanceusa.com/blog/how-nist-can-protect-the-cia-triad-including-the-often-overlooked-i-integrity. Accessed 17 Dec 2019
22. SNIA: Contact us via LiveChat!. SNIA (November 2018)
23. Schopmeyer, K.: Automation of SMI-S managed storage systems with Pywbem, p. 47 (2017)
24. Hubbert, S.: Datacenter storage; cost-effective strategies, implementation, and management. SNIA (2011)
25. Daniel, E., Vasanthi, N.A.: LDAP: a lightweight deduplication and auditing protocol for secure data storage in cloud environment. Cluster Comput. **22**(1), 1247–1258 (2017). https://doi.org/10.1007/s10586-017-1382-6
26. Park, S.-W., Lim, J., Kim, J.N.: A secure storage system for sensitive data protection based on mobile virtualization. Int. J. Distrib. Sens. Netw. **11**(2), 929380 (2015). https://doi.org/10.1155/2015/929380

27. Butler, K.R.B., McLaughlin, S., McDaniel, P.D.: Rootkit-resistant disks. In: Proceedings of the 15th ACM Conference on Computer and Communications Security - CCS 2008, Alexandria, Virginia, USA, p. 403 (2008). https://doi.org/10.1145/1455770.1455821
28. Tang, Y., et al.: NodeMerge: template based efficient data reduction for big-data causality analysis. In: Proceedings of the 2018 ACM SIGSAC Conference on Computer and Communications Security - CCS 2018, Toronto, Canada, pp. 1324–1337 (2018). https://doi.org/10.1145/3243734.3243763
29. Hibbard, E.: Intro to encryption and key management: why, what and where? SNIA (2016)
30. Willett, M.: Implementing stored-data encryption, p. 50 (2012)
31. SNIA: Cloud data management interface (CDMI^{TM}) version 1.1.1. SNIA (March 2015)
32. SNIA: Storage security: an overview as applied to storage management version 1. SNIA (August 2016)
33. SNIA: Storage security: encryption and key management. SNIA (August 2015)
34. McKay, K.A., Polk, W.T., Chokhani, S.: Guidelines for the selection, configuration, and use of transport layer security (TLS) implementations. NIST (April 2014)
35. ENISA: ENISA threat landscape report 2018 15 top cyberthreats and trends. ENISA (January 2019)
36. Zhou, L., Varadharajan, V., Gopinath, K.: A secure role-based cloud storage system for encrypted patient-centric health records. Comput. J. **59**(11), 1593–1611 (2016). https://doi.org/10.1093/comjnl/bxw019
37. Sarkar, M.K., Chatterjee, T.: Enhancing data storage security in cloud computing through steganography (2014)
38. Krahn, R., Trach, B., Vahldiek-Oberwagner, A., Knauth, T., Bhatotia, P., Fetzer, C.: Pesos: policy enhanced secure object store. In: Proceedings of the Thirteenth EuroSys Conference on - EuroSys 2018, Porto, Portugal, pp. 1–17 (2018). https://doi.org/10.1145/3190508.3190518
39. Hibbard, E.A.: Best practices for cloud security and privacy. SBIA (2014)
40. Hibbard, E., Rivera, T.: Reforming EU data protections… No ordinary sequel. SNIA (September 2014)
41. SNIA: TLS specification for storage systems. SNIA (November 2014)
42. SNIA: Storage networking industry association. SNIA (March 2018)
43. PCI SSC: Payment card industry (PCI) data security standard report on compliance. PCI DSS v3.2 Template for Report on Compliance. PCI (June 2018)
44. PCI SSC: PCI DSS quick reference guide understanding the payment card industry data security standard version 3.2. PCI SSC (October 2010)
45. Schaffer, K.: ITL bulletin May 2019 FIPS 140-3 adopts ISO/IEC standards. NIST, p. 3 (May 2019)
46. Brandão, L., Davidson, M., Mouha, N., Vassilev, A.: ITL bulletin for APRIL 2019 time to standardize threshold schemes for cryptographic primitives. NIST, p. 6 (April 2019)
47. Zyskind, G., Nathan, O., Pentland, A.: Decentralizing privacy: using blockchain to protect personal data. IEEE (Juk 2015)
48. Wang, H., Yang, D., Duan, N., Guo, Y., Zhang, L.: Medusa: blockchain powered log storage system. IEEE (March 2019)
49. Meslhy, E., Abd elkader, H., Eletriby, S.: Data security model for cloud computing. J. Commun. Comput. **10**, 1047–1062 (2013). https://doi.org/10.13140/2.1.2064.4489
50. Jovanovic, V., Mirzoev, T.: Teaching storage infrastructure management and security. In: 2010 Information Security Curriculum Development Conference, New York, NY, USA, pp. 41–44 (2010). https://doi.org/10.1145/1940941.1940952

51. Butler, K.R.B., McLaughlin, S.E., McDaniel, P.D.: Non-volatile memory and disks: avenues for policy architectures. In: Proceedings of the 2007 ACM Workshop on Computer Security Architecture, New York, NY, USA, pp. 77–84 (2007). https://doi.org/10.1145/1314466.1314479
52. Paik, J.-Y., Choi, J.-H., Jin, R., Wang, J., Cho, E.-S.: A storage-level detection mechanism against crypto-ransomware, pp. 2258–2260 (2018). https://doi.org/10.1145/3243734.3278491
53. Hibbard, E.A.: SNIA storage security best practices. SNIA (2015)
54. Hibbard, E.A.: SNIA storage security best practices. SNIA (2011)
55. ISO: ISO/IEC 27040:2015 information technology—security techniques—storage security. ISO (2015)
56. SNIA: NVM programming model (NPM). SNIA (June 2017)
57. SNIA: Storage security: fibre channel security. SNIA (2016)
58. SNIA: Sanitization. SNIA (March 2015)
59. SNIA: Architectural model for data integrity. SNIA (March 2012)
60. SNIA: Common RAID disk data format specification. SBIA (March 2009)
61. SNIA: Hypervisor storage interfaces for storage optimization white paper. SNIA (June 2010)
62. Gordan, J.: Practical Data Security (Unicom Applied Information Technology), 1st edn. (2019)
63. Talib, A.M., Atan, R., Murad, M.A.A., Abdullah, R.: A framework of multi agent system to facilitate security of cloud data storage. In: International Conference on Cloud Computing Virtualization, pp. 241–258 (2010)
64. Dharma, R., Venugopal, V., Sake, S., Dinh, V.: Building secure SANs. EMC (April 2013)
65. IBM: IBM storage insights: security guide. IBM (September 2019)
66. Hasan, R., Yurcik, W.: A statistical analysis of disclosed storage security breaches. In: Proceedings of the Second ACM Workshop on Storage Security and Survivability, New York, NY, USA, pp. 1–8 (2006). https://doi.org/10.1145/1179559.1179561
67. Vasilopoulos, D., Elkhiyaoui, K., Molva, R., Onen, M.: POROS: proof of data reliability for outsourced storage. In: Proceedings of the 6th International Workshop on Security in Cloud Computing, New York, NY, USA, pp. 27–37 (2018). https://doi.org/10.1145/3201595.3201600
68. Zhu, Y., Wang, H., Hu, Z., Ahn, G., Hu, H., Yau, S.S.: Dynamic audit services for integrity verification of outsourced storage in clouds. In: 2011 Proceedings of ACM Symposium on Applied Computing (SAC), pp. 1550–1557 (December 2010)
69. Subha, T., Jayashri, S.: Efficient privacy preserving integrity checking model for cloud data storage security. IEEE (January 2017)
70. Hou, H., Yu, J., Hao, R.: Cloud storage auditing with deduplication supporting different security levels according to data popularity. ScienceDirect (Nay 2019)
71. Schulz, G.: Cloud and Virtual Data Storage Networking, 1st edn. CRC Press, Boca Raton (2011)
72. Kwon, J., Johnson, M.E.: Meaningful healthcare security: does 'meaningful-use' attestation improve information security performance? EBSCOhost (December 2018)
73. Dell EMC: Dell EMC Unity^{TM} family security configuration guide. Dell EMC (December 2018)
74. HDS: Hitachi virtual storage platform (VSP) encryption engine non-proprietary Cryptographic-FIPS 140-2 Module Security Policy. HDS (February 2019)
75. Li, L., Qian, K., Chen, Q., Hasan, R., Shao, G.: Developing hands-on labware for emerging database security. In: Proceedings of the 17th Annual Conference on Information Technology Education, New York, NY, USA, pp. 60–64 (2016). https://doi.org/10.1145/2978192.2978225

76. SNIA: Linear tape file system (LTFS) format specification. SNIA (May 2019)
77. SNIA: Self-contained information retention format (SIRF) specification. SNIA (December 2016)
78. SNIA: Multipath management API. SNIA (March 2010)
79. McMinn, M.: Information management—extensible access method (XAM)—Part 1: architecture. SNIA (June 2009)
80. McMinn, M.: Information management – extensible access method (XAM) – Part 2: C API. SNIA (June 2009)
81. McMinn, M.: Information management – extensible access method (XAM) – Part 3: Java API. SNIA (June 2009)
82. Dutch, M.: A data protection taxonomy. SNIA (June 2010)
83. SNIA: Data protection best practices. SNIA (October 2017)
84. Schopmeyer, A., Somasundaram, G.: Information Storage and Management: Storing, Managing, and Protecting Digital Information. O'Reilly, Sebastopol (2009)
85. Chang, Z., Hao, Y.: The research of disaster recovery about the network storage system base on 'Safety Zone.' IEEE (October 2009)
86. Jian-hua, Z., Nan, Z.: Cloud computing-based data storage and disaster recovery. IEEE (August 2011)
87. Wang, X., Cheng, G.: Design and implementation of universal city disaster recovery platform. IEEE (May 2018)
88. Bollinger, J., Enright, B., Valite, M.: Crafting the InfoSec Playbook: Security Monitoring and Incident Response Master Plan, 1st edn. O'Reilly, Sebastopol (2015)
89. Zhou, J.: On the security of cloud data storage and sharing. In: Proceedings of the 2nd International Workshop on Security in Cloud Computing, New York, NY, USA, pp. 1–2 (2014). https://doi.org/10.1145/2600075.2600087
90. Chen, M., Zadok, E.: Kurma: secure geo-distributed multi-cloud storage gateways. In: Proceedings of the 12th ACM International Conference on Systems and Storage - SYSTOR 2019, Haifa, Israel, pp. 109–120 (2019). https://doi.org/10.1145/3319647.3325830
91. Carlson, M., Espy, J.: IP-based drive management specification. SNIA (January 2017)
92. SNIA: iSCSI management API. SBIA (June 2008)
93. Fuxi, G., Yang, W.: Data Storage at the Nanoscale, 1st edn. Jenny Stanford Publishing (2015)
94. Rouse, M.: What is blockchain storage? SearchStorage (June 2019). https://searchstorage.techtarget.com/definition/blockchain-storage. Accessed 15 Dec 2019
95. BlockApps: How blockchain will disrupt data storage. BlockApps (Dec 2017). https://blockapps.net/blockchain-disrupt-data-storage/. Accessed 15 Dec 2019
96. Xu, Y.: Section-blockchain: a storage reduced blockchain protocol, the foundation of an autotrophic decentralized storage architecture. IEEE (December 2018)
97. Zheng, Q., Li, Y., Chen, P., Dong, X.: An innovative IPFS-based storage model for blockchain. IEEE (December 2018)
98. Veleva, P.: Personal data security for smart systems and devises with remote access. EBSCOhost (2019)
99. Virtualization Market Now 'Mature,' Gartner Finds: InformationWeek. https://www.informationweek.com/cloud/infrastructure-as-a-service/virtualization-market-now-mature-gartner-finds/d/d-id/1325529. Accessed 13 Dec 2019
100. Reasons Behind Storage Virtualization Software Use 2015: Statista. https://www.statista.com/statistics/678925/worldwide-storage-virtualization-software-use-reasons/. Accessed 13 Dec 2019
101. Liu, S.: Global data storage problems 2016–2017. Statista. https://www.statista.com/statistics/752840/worldwide-data-storage-problems/. Accessed 13 Dec 2019
102. ENISA Threat Landscape Report 2018. https://www.enisa.europa.eu/publications/enisa-threat-landscape-report-2018. Accessed 13 Dec 2019

103. DiGiacomo, J.: Active vs passive cyber attacks explained. Revision Legal, 14 February 2017. https://revisionlegal.com/cyber-security/active-passive-cyber-attacks-explained/. Accessed 22 Dec 2019
104. CMS: HIPAA basics for providers: privacy, security, and breach notification rules. CMS (September 2018)
105. U.S. Department of Health & Human Services - Office for Civil Rights. https://ocrportal.hhs.gov/ocr/breach/breach_report.jsf. Accessed 21 Dec 2019
106. O'Neil, C.: Opinion: big-data algorithms are manipulating us all. Wired, 18 October 2016

Taming the Digital Bandits: An Analysis of Digital Bank Heists and a System for Detecting Fake Messages in Electronic Funds Transfer

Yasser Karim(✉) and Ragib Hasan(✉)

Department of Computer Science, University of Alabama at Birmingham, Birmingham, AL, USA
{yasser,ragib}@uab.edu

Abstract. In recent years, financial crimes and large scale heists involving the banking sector have significantly increased. Banks and Financial Institutions form the economic and commercial backbone of a country. An essential function of banks is the transfer of funds domestically or internationally. Most banks today transfer money by using electronic fund transfer systems such as the Automated Clearing House (ACH) or messaging systems such as SWIFT, FedWire, Ripple, etc. However, vulnerabilities in the use of such systems expose banks to digital heists. For example, the 2016 heist in the central bank of Bangladesh used the SWIFT network to send fake messages. It almost resulted in the theft of nearly $1 billion, which is one-sixth of the total foreign currency reserve of Bangladesh. Similar attacks have happened in many other countries as well. In this paper, we discussed multiple such incidents. From those incidents, we systematically analyze two such events – the Bangladesh Bank heist and the DNS takeover of Brazilian banks – to understand the nature and characteristics of such attacks. Through our analysis, we identify common and critical security flaws in the current banking and messaging infrastructures and develop the desired security properties of an electronic funds transfer system.

1 Introduction

Banks and other financial institutions form the necessary foundation of a monetary system. Unfortunately, such institutions are also subject to frequent attacks and manipulations by criminals motivated by financial gains. With the advent of the Internet, banks and financial institutions have embraced electronic communication systems and information technology for their fund transfer operations. However, the criminals are now frequently resorting to attacking such communication systems to defraud banks. Different types of online criminal activities such as hacking, data breach, phishing, ransomware, malware attack, etc. targeting financial institutions have increased in recent years. According to a 2018 report from the US Federal Bureau of Investigation (FBI), the number of criminal incidents related to banks (robberies and/or burglaries) in 2018 was around four

K.-K. R. Choo et al. (Eds.): NCS 2020, AISC 1271, pp. 193–210, 2021.
https://doi.org/10.1007/978-3-030-58703-1_12

thousand (Table 1) [29]. On the other hand, a report from the US Department of Treasury (Table 2) shows that in 2018, 154,000 out of 394,000 reported incidents (40%) were related to online/internet banking. Also, according to a report by the Internet Crime Complaint Center (IC3), approximately $122 million was lost to banking-related crimes such as credit card fraud, personal/corporate data breach, etc. in 2018 (Fig. 1). The total loss in 2015 for Internet-related crimes was approximately $1 billion [29].

In recent years, the scale of bank heists by cybercriminals has significantly increased. In February 2016 [12], $81 million was lost from the foreign reserve of the Bangladesh Bank – the central bank of Bangladesh in a heist that spanned at least four countries and raised questions about the international banking systems' vulnerability against cyber-attacks. The perpetrators were able to issue multiple instructions via SWIFT – a popular bank messing network, in an attempt to steal $1 billion from Bangladesh Bank. Five transactions to withdraw approximately $101 million from a Bangladesh Bank account at the Federal Reserve Bank of New York were successful. The economy of a developing country such as Bangladesh immensely depends on its foreign reserve. Therefore, such heists can be debilitating to the economy of the whole country. In another incident in 2015, Ecuador's Banco del Austro lost $12 million in a similar attack [14]. In both events, the computer networks of the banks were compromised, and hackers were able to gain access to the SWIFT network. Such incidents indicate the importance of securing the messaging networks of banks. It is vital to examine digital bank heists to understand the common patterns and techniques used and to develop systems that can prevent such heists.

Table 1. FBI report on Bank related crimes 2018 [29]

	Robberies	Burglaries	Larcenies
Commercial Banks	2,707	47	2
Mutual Savings Banks	5	0	0
Savings and Loan Associations	23	0	0
Credit Unitos	215	6	0
Armored Carrier Companies	10	1	0
Others	15	2	0
Totals:	2,975	56	2

In this paper, we examine several digital bank heists to explore the nature and modus operandi of such crimes. We identify the security flaws that are frequently exploited by the perpetrators to launch such attacks. Our analysis indicates that, in each documented case where the SWIFT network was used to issue a fake message, the criminals followed three basic patterns. They are: **1)** Attackers used malware to circumvent a bank's local security systems; **2)** They gained access to the SWIFT messaging network, and **3)** Fraudulent messages were sent via SWIFT to initiate cash transfers from accounts at larger banks.

Contribution: The contributions of this paper are as follows:

1. We have investigated and categorized multiple online/internet bank heist incidents and discovered common attack patterns that are used in digital heists – to the best of our knowledge, this is the first academic study of such incidents;
2. We have conducted a comprehensive security analysis of two major bank heist incidents and identified key security flaws in banking systems that were exploited;

Table 2. Depository institution related crime incident 2018 [40])

Suspicious activity	Filing count
Automated clearing house	37,338
Check	1,263
Consumer loan	71,548
Credit/debit card	87,364
Healthcare	687
Mail	8,358
Mass-marking	3,509
Pyramid scheme	307
Wire transfer	29,298
Other	42,199
Totals:	394,000

Organization: The paper is organized as follows: Sect. 2 provides a brief overview of electronic fund transfer systems and the architecture and mechanism of SWIFT. In Sect. 3, we analyze and classify different types of bank heist incidents. In Sect. 4, we examine two case studies involving attacks on Bangladeshi and Brazilian banks. We discussed related work in Sect. 5, and we concluded in Sect. 6.

2 Background

To understand the different steps of a digital bank heist, we first have to know how electronic money transfer happens. In this section, we explore different electronic money transfer techniques.

2.1 Electronic Fund Transfer

Electronic funds transfer (EFT) is the electronic transfer of money from one bank account to another, either within a single financial institution or across multiple institutions, via computer-based systems, without the direct intervention of bank staff [4]. There are three main electronic methods of transferring money: Automated Clearing House Transfers, Wire Transfers, and Electronic Transfers via third-party systems [2].

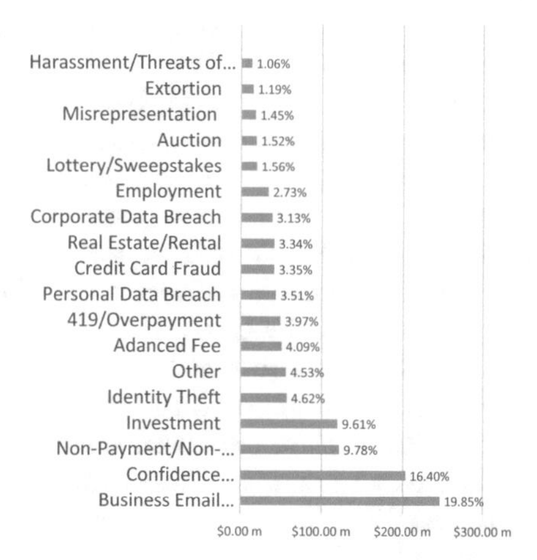

Fig. 1. Internet crime by monetary loss 2018

Automated Clearing House Transfers: The Automated Clearing House (ACH) is an electronic network used by financial institutions to process transactions in batches. Both credit and debit transactions are processed via ACH. The credit transfers include deposit, payroll, vendor payments, and point of sales.

The debit transfers include consumer payments on insurance, mortgage loans and other kinds of bills. In 2015, this network processed nearly 24 billion ACH transactions with a total value of $41.6 trillion [1].

Wire Transfer: A wire transfer is an electronic payment service for transferring funds by wire, for example, through SWIFT, the Federal Reserve Wire Network (FedWire), or the Clearing House Interbank Payments System [4]. Unlike the batch-processing nature of ACH transfers, wire transfers are designed for individual transactions. The most significant benefit of wire transfers is the speed or availability of funds.

Electronic Transfers via Third-Party Systems: Apart from ACH and Wire Transfer system, there are third party systems that can be used to transfer money. These systems obfuscate the whole banking transaction process and provide a natural mode of access for users. An example of such a system is PayPal [9].

2.2 Architecture of SWIFT System

SWIFT, or the Society for Worldwide Interbank Financial Telecommunication, is the world's most extensive electronic payment messaging system, facilitating the exchange of more than $8 trillion a day, according to 2018 estimates [15].

SWIFT is a centralized system which allows storing and forwarding mechanism along with transaction management (Fig. ??). For example, if bank A wants to send a message to bank B with authorization from an institution C, bank A formats the message following the specification provided by SWIFT and sends it to the SWIFT through a separate secure network. The SWIFT system assures that it is secure and reliable delivery to B after the proper action by C. SWIFT messages are structured in a format known as FIN. It uses a system of codes to indicate the source, destination, and mechanism of a transfer request. These codes are alphanumeric. As SWIFT does not transfer the money, institutions that use the network also need banking or financial relationship to move funds. Each institution usually installs a separate dedicated interface for SWIFT. Users can log in to SWIFT terminals to manually enter messages. Messages can also be generated by the institution's computer system through a secure API and transferred to the terminal. The terminal then sends the SWIFT message to the SWIFT system. At first, this message is sent to the regional processors in the sender's country. From there, the regional processor forwards the message to operating center after proper checking and storing the message. Then the operating message passes the message on to the regional processor in the recipient's country. That processor

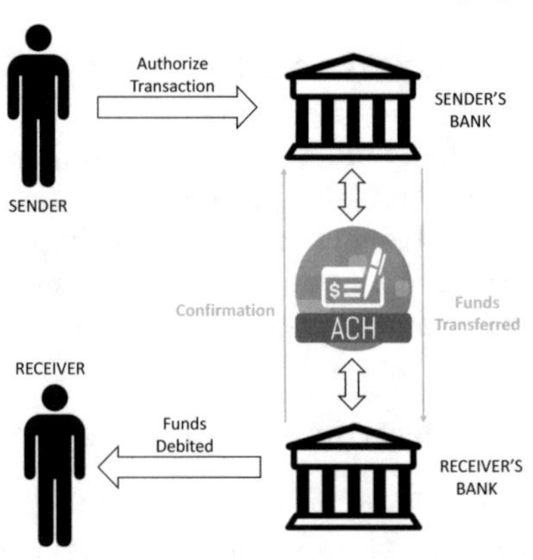

Fig. 2. Automated clearing house

transfers the message to the receiver's terminal, and then a confirmation is sent back to the operating center, which also is forwarded to the sender's terminal. Later, these transactions are required to be audited (Figs. 2 and 3).

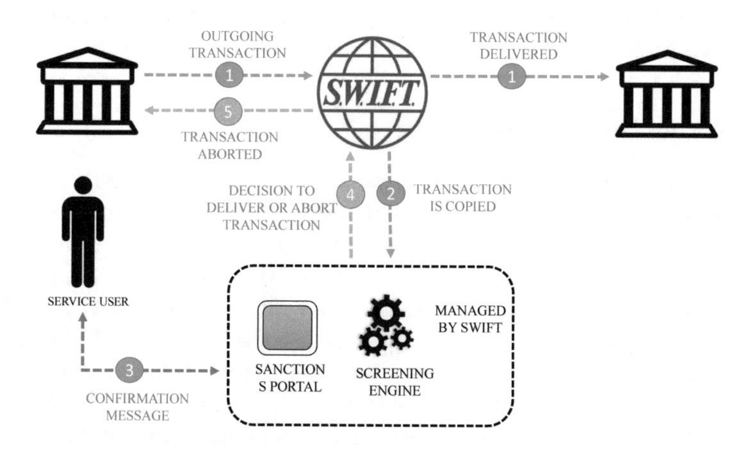

Fig. 3. Mechanism of SWIFT network

3 Classification of Digital Bank Heists

We collected digital bank heist incidents reported in the media between 2010 and 2018. Our analysis suggests that we can classify digital bank heists into the following three broad classes: **1)** Data Breach & Financial Account Hack. **2)** Banking Service Disruption. **3)** Compromised Bank Messaging Network.

3.1 Data Breach & Financial Account Hack

Financial information is confidential, and Banks, Credit Unions are the prime hubs for this information. In the past, several times, perpetrators were able to breach banking systems and leaked valuable information about the account holders. In October 2014 [17], in an SEC filing, it was reported that the private data of 70 million households and 7 million small business might have compromised from JPMorgan. Despite the bank spending $250 million annually on cybersecurity [5], this leak cost approximately $1 billion. In 2012, 1.5 million card accounts of Global Payments were compromised. It was reported that this breach cost more than $90 million. In 2011, 360,000 credit card holders of Citibank were affected by data breach [3]. It was estimated that the breach cost $19 million. In 2016 [16], HDFC, ICICI, State Bank of India, and several other banks from India suffered a large scale debit card breach attack. Almost 2.6 million debit cards were compromised. The attacker exploited a bug in Hitachi card reader and stole information of debit cards, most of them were on VISA and Master Card platforms. Also 600,000 cards of RuPay system were also breached [11].

In this type of criminal incidents, bank or credit card accounts are compromised to steal money from the user. Using malware, phishing, scamming, etc. cyber criminals get hold of financial accounts. Then these accounts are used to issue money transfer or illegitimate order in online shopping. In 2016 [20], almost 20,000 accounts of Tesco Bank were hacked and money was stolen from the accounts. Approximately 2.5 million pounds was lost.

3.2 Banking Service Disruption

Banks are often a target for DDoS attacks. For example, in 2017, the Lloyds Banking Group suffered a 48-h denial of service attack [18]. Approximately 20 million accounts in UK were blocked and the users were unable to access them. Another incident that was published in one of the reports by Kaspersky Lab [13] that the attacker hijacked a bank's entire internet footprint. The attacker changed the Domain Name System registrations of all 36 of the Brazilian bank's online properties [13]. As a result, the user of the bank's website and mobile app were re-routed to spoofed sites where users' information was compromised. In 2016, five Russian banks suffered days-long DDoS attacks [19]. On September 19, 2012, the websites of Bank of America, JPMorgan Chase, Wells Fargo, U.S. Bank, and PNC Bank suffered day-long slowdowns and were sporadically unreachable to customers [22] (Table 3).

3.3 Compromised Bank Messaging Network

Every bank has collaborations with other banks for transferring funds among them. To request a funds transfer, most of the financial institutions use a

Table 3. Notable cyber crimes relating banking organization (2011–2018)

Time	Victim	Type	Description	Loss
Feb 2016	Central Bank of Bangladesh	Compromised Messaging Network (SWIFT)	The perpetrators gained access to SWIFT messaging system by using malware and issued fund transfer messages	$81 million
Jan 2015	Banco del Austro, Ecuador	Compromised Messaging Network (SWIFT)	The perpetrators gained access to SWIFT messaging system by using malware and issued fund transfer messages	$12 million
Dec 2015	Tien Phong Bank, Vietnam	Compromised Messaging Network (SWIFT)	The perpetrators gained access to SWIFT messaging system by using malware and issued fund transfer messages	0
Oct 2014	JPMorgan, USA	Data Breach	Approximately 76 million household information was stolen	Uncertain
Oct 2016	Multiple Brazilian Banks	Banking Service Disruption	The Internet footprint of multiple banks were compromised by changing the domain name system registration	Uncertain

messaging network such as SWIFT or Fedwire [6], to issue transfer requests. In several notable incidents, criminals got access to machines where messaging service was configured and then sent messages to other banks to transfer money. In February 2016, $81 million was stolen from the account of Bangladesh Bank in the Federal Reserve Bank of New York in a similar way [12]. Before that, in 2013, another bank of Bangladesh named Sonali Bank was the target for the hackers. The adversaries stole $250,000 with fraudulent swift messaging. Also, $10 million from a Ukrainian bank [21] was removed by issuing a SWIFT message via the compromised local network. The first reported incident of this kind was theft $12 million from Banco del Austro in Ecuador [14] in January 2015. In the same year, Vietnam's Tien Phong Bank reported an unsuccessful attack of this kind [23].

4 Case Studies

In this section, we examine two case studies involving large scale attacks on the Bangladesh Bank and a Brazilian Bank. We examine both incidents in great detail to identify the techniques used (Fig. 4).

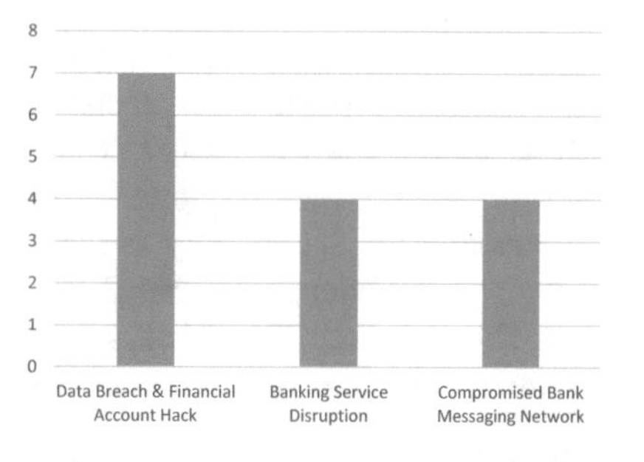

Fig. 4. Major bank related online criminal incidents between 2011 and 2018

4.1 Incident I: Central Bank of Bangladesh

In February 2016, Bangladesh Bank was attacked by unknown criminals, and an attempt was made to steal money from the foreign currency reserves (deposited at the Federal Reserve Bank of New York).

Target: Attackers identified weaknesses in the security of Bangladesh Bank's computer network. Leveraging the weaknesses, attackers attempted to steal $951 million from the Bangladesh Bank's foreign currency reserve account at the Federal Reserve Bank of New York. The heist took place between February 4 and 5 when Bangladesh Bank's offices were closed due to the weekend [12].

Scenario: Attackers first broke into Bangladesh Bank's computer network and identified the network structure. Using sniffers and other exploit tools, the attackers learned the funds transfer protocol. Later, the attackers managed to compromise the credentials of persons authorized to make payment requests. During the heist, attackers sent 36 funds transfer requests to the Federal Reserve Bank of New York to transfer funds from Bangladesh Bank's account there to accounts in Sri Lanka and the Philippines. One of the transfer messages had spelling mistakes, which raised suspicion. So after six transactions, the Federal Reserve Bank of New York blocked the rest of the transactions [8].

Tools: It was identified later that the Dridex malware was used for the attack. It is spread through e-mail that infiltrates computers and harvest information like user names and passwords, which are used to gain access to privileged networks. Dridex was first identified by security researchers in 2014 [12].

Synopsis: Banking networks are highly secure environments with a variety of unique internal processes, software, and systems. So the attacker maybe with the help of insider or using conventional phishing or trojan horse tries to compromise some computer inside the banking network. Once some bank employee's computer is infected, the attacker patiently listened and learned. For example, experts believe that in Bangladesh Bank case, the attackers observed employees' activities for a least one month. Malware like Carbanak can record the employee's desktop display and sent the video to the remote attacker. In addition to the standard malware behavior of capturing user credentials, this video of desktop display enabled the attacker to watch an employee and learn the internal processes of the bank. Researchers also found from the sample malware of Bangladesh Bank attack that the attackers planted list of SWIFT codes of several destination bank accounts. So once they had done enough reconnaissance of the network they placed another malware to observe SWIFT network messages. The malware was also monitoring the SWIFT message to certain destination banks to learn the pattern of regular transactions. The attacker issued the message on a Thursday night, which is the weekend in Bangladesh. So the next day, the Federal Reserve Bank of New York approved five transactions, and when they blocked other transactions because of spelling mistakes, they could not verify with the employees of Bangladesh Bank.

Aftermath: Although the Federal Reserve Bank of New York was able to block thirty transactions, five transactions worth $101 were passed through. Among them, $20 million deposited in a Srilankan bank account were recovered later. The rest of the $81 million were deposited in five accounts of Rizal Commercial Bank Corporation in the Philippines. From those accounts, the money was

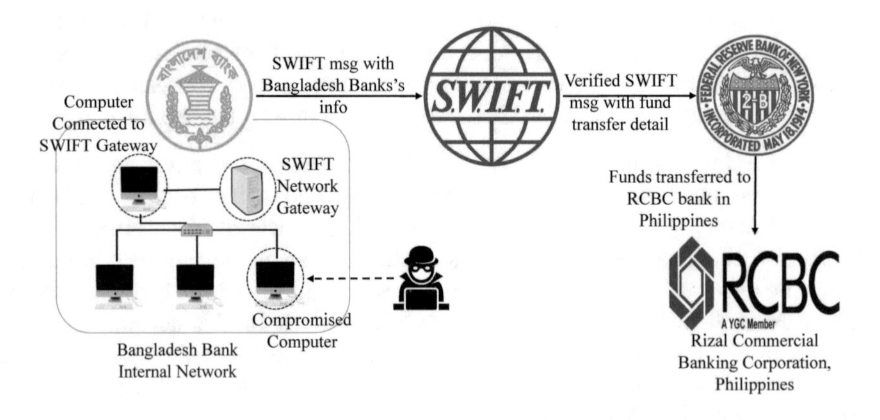

Fig. 5. Bangladesh bank heist: event flow

transferred to casino accounts and some later transferred to another casino in Hong Kong. After that the money trail was lost. After this incident, the chief of Bangladesh Bank resigned from his position (Fig. 5).

Investigation: After the incident, the Bangladesh government initiated the initial investigation. In their investigation, it was discovered that the malware was installed in January 2016 [7]. The Philippines' National Bureau of Investigation (NBI) also launched a probe to investigate the part o Filipino involvement. On August 12, 2016, RCBC had to pay 1 billion Filipino dollars as a penalty to Central Bank of the Philippines [10]. The United States also launched separate investigation. FireEye's Mandiant forensics division and World Informatix Cyber Security, two US-based companies, were hired by different nations and the US state department for the investigation. From their reports, the perpetrators' were familiar with the internal procedures of Bangladesh Bank. In a separate report, the US Federal Bureau of Investigation (FBI) says that agents have found evidence pointing to at least one bank employee acting as an accomplice. With evidence looking to several more people as possibly assisting hackers in navigating the Bangladesh Bank's computer system.

Security Analysis: Now we go in depth of this incident to find out the weak security specks which favored the attacker.

Isolation of SWIFT Network: In most financial institutions, the SWIFT network is configured separately from other networks of that institution. Also, the machines that are connected to the SWIFT network are kept isolated from other networks. In the case of Bangladesh Bank, the perpetrators were able to put malware on the SWIFT network-connected machine. The SWIFT interface allows manually entering the message as well as auto-generated messages from the institution's computer system. In this incident, the malware was generating the messages, and the SWIFT network was not aware of that.

Confirmation of a SWIFT Transaction: In both cases of Bangladesh Bank and Banco del Austro in Ecuador, the confirmation messages were interrupted by the attackers. After the malware sent the SWIFT messages that stole the funds, it deleted the database record of the transfers then took further steps to prevent confirmation messages from revealing the theft. In the Bangladesh Bank incident, the confirmation messages would have appeared on a paper report; the malware altered the paper reports when they were sent to the printer. In the second case, the bank used a PDF report; the malware altered the PDF viewer to hide the transfers.

Disputation of a SWIFT Message: After the incident, the authority of Bangladesh Bank blamed the Federal Reserve Bank of New York that the transactions should not be allowed by them. Their first response was that the Federal Reserve Bank of New York was hacked. But after the investigation by the New York reserve bank, it was confirmed that the messages were validated by the SWIFT system. No evidence of any hacking was found. Further investigation also showed that the malware was installed on one of Bangladesh Bank's employee's computers.

Internal Security of Financial Institutions: The security of banking systems, especially in the third world or least developed countries, are not sophisticated enough. In both cases, the perpetrators managed to acquire a valid username and password for the SWIFT network by observing banks' internal network. These banks were not well prepared for persistent attacks, and their internal networks had security niches that were exploited by the attackers.

The Role of an Insider: From this incident, it seems that the attacker had eyes on the inside, either within the bank or with SWIFT. The level of insight they had into the SWIFT platform raises a concern about the possibility of insiders. This is highly plausible that one or multiple insiders were involved in those heists.

Persistent Attacker: The Bangladesh Bank incident revealed that the attacker had adequate knowledge about the banking networks. From investigation, it was found that the malware was installed in January, and the attacker observed banking activity at least for a month. They chose the date for launching the attack in such a precise manner that it fell on weekends and followed by the Chinese new year celebration. It shows the sophistication of the attack.

4.2 Incident II: Brazilian Banks

Now we look into the incident of Brazilian Banks [13]. This incident exposed a new way to disrupt banking service. In this incident, hackers were able to hijack a bank's entire Internet footprint.

Target: Hackers changed the domain name system registration of 36 banks' online properties. As a result, users were directed towards phishing sites by banks' websites and mobile apps. In practice, that meant the hackers could steal login credentials at sites hosted at the bank's legitimate web addresses. Kaspersky researchers believe the hackers may have even simultaneously redirected all transactions at ATMs or point-of-sale systems to their servers, collecting the credit card details of anyone who used their cards [13].

Scenario: The attackers compromised the bank's account at Registro.br. That's the domain registration service of NIC.br, the registrar for sites ending in the Brazilian.br top-level domain, which they say also managed the DNS for the bank. With that access, the researchers believe, the attackers were able to change the registration simultaneously for all of the bank's domains, redirecting them to servers the attackershad set up on Google's Cloud Platform.

Tools: It was not disclosed what kind of tools were used by the attackers.

Synopsis: With that domain hijacking in place, anyone visiting the bank's website URLs was redirected to lookalike sites. And those sites even had valid HTTPS certificates issued in the name of the bank, so that visitors' browsers would show a green lock and the bank's name, just as they would with the real sites. It was also found that the certificates had been issued six months earlier by Let's Encrypt, the non-profit certificate authority that's made obtaining an HTTPS certificate easier in the hopes of increasing HTTPS adoption.

Also, the hackers were able to shut down the banks' email sending shutdown. Aside from mere phishing, the spoofed sites also infected victims with a malware that disguised itself as an update to the trustier browser security plug-in that the Brazilian bank offered customers. According to Kaspersky's analysis, the malware harvests, not just banking logins—from the Brazilian banks as well as eight others—but also email and FTP credentials, as well as contact lists from Outlook and Exchange, all of which went to a command-and-control server hosted in Canada.

Aftermath: After around five hours, the bank regained control of its domains, likely by calling to NIC.br and convincing it to correct the DNS registration. How many banks' customers were caught up in the DNS attack was not disclosed by the banks. But the researchers believe that the perpetrators could have infected millions of customers' account detail not only from their phishing scheme and malware but also from redirecting ATM and point-of-sale transactions to infrastructure they controlled. Although NIC.br, the domain name registration organization, acknowledged some vulnerabilities in their websites, they denied all the complaints about being hacked.

Security Analysis

Domain Registration for Financial Institutions: The case of Brazilian Banks showed us another way to compromise banking service without attacking the banking network directly. The attackers managed to compromise the Domain Registration organization for a financial institution and changed their record, which resulted in directing legit users to spoofed websites.

Certificate Authority: In the Brazilian incident, the spoofed websites were look-alike of original websites, and those sites had valid certificates, which caused the browser to show a green lock sign. As a result, a lot of users trusted those sites and put their valuable information into those spoofed sites. The certificate authority made their issuing policies easy to promote HTTPS adoption. But once someone gained control of DNS, their policies were failed to identify them as malicious. So the certificate authority should be more aware and cautious about issuing a certificate and should contain policies regarding this kind of incident.

5 SecureSWIFT

We proposed a cloud-based service framework SecureSWIFT that can be used to establish secure communication between two or multiple banks. Our framework provides a separate messaging network that will automatically alert clients based on delivered SWIFT messages. Using state of the art machine learning techniques, the system will detect suspicious and malicious SWIFT messages which are irregular in nature in comparison with previous valid messages. After detection, the system will alert both clients and flagged that message as risky. The main idea is to build a separate system that will provide a platform to detect and alert suspicious SWIFT messages without hampering the SWIFT network.

5.1 System Architecture

Our proposed service framework has following components (Fig. 6):

- Access & Management Interface
- Registration & Key Management
- User Profiles & Validation
- Message Transfer Point
- Machine Learning Engine
- Message Archive
- Alert System
- Message Gateway
- Messages Reporting.

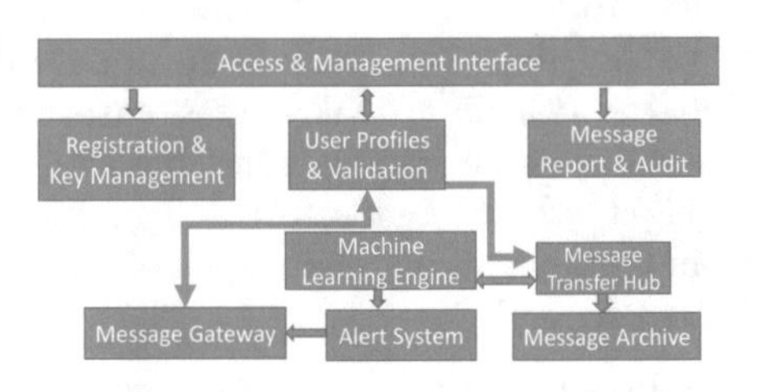

Fig. 6. System architecture

5.2 Operational Model

Now we will describe the operation model step by step.

Registration and Key Assignment: The first and foremost step before using the service is to register and to acquire the key for all kinds of communication. A user submits her request for an account and the key with proper information. After proper check through, the user account is created, and a pair of the public-private key is assigned to the user. The private key is removed from the system, and the public key is kept in the user profile. Also, this is kept in the global key database so that other users can use this public key to send a secure message to that user.

Setting Up Message Transfer Mechanism: The next step in the process is setting up the SWIFT message transferring system and process. The user could set up a real-time system to transfer the SWIFT message to the cloud, or they can send them in a batch. This process can be automated or manual depending on the option chosen by the user. All the transfer process is handled through the Message Transfer System. The messages that are transferred to the cloud should be with a digital signature with that user's private key so that each message can be verified with the public key.

Alert System Preferences: The next step is providing the preferences of the alert system. The user needs to set up her alert system. Both email and SMS services can be enabled. Also, upon receiving a suspicious SWIFT message, the system can send a message to both parties if both users are using the service. The user can also enable this option.

ML Algorithm Selection: The next important thing is the selection of a machine learning algorithm. The engine will provide several options to select. Also, the user needs to mention how often the training process will take place. They can also select how much data will be used in the training process. Also, the system will provide suggestions to the user about this critical information.

Message Detection: After all the configuration and setting up are done, the system is ready for identifying spoofed messages. The message transfer system continuously receives SWIFT messages. It transfers each message to the recommendation model of the machine learning engine. If any message detected as risky, it would be immediately transferred to the flagged message database. Also, the model is trained with the updated database. As it is a machine learning-based approach, there is always a chance of false positives. So this re-training will increase the accuracy of the system. Besides, the reporting system will allow the user to correct wrongly places messages (Fig. 7).

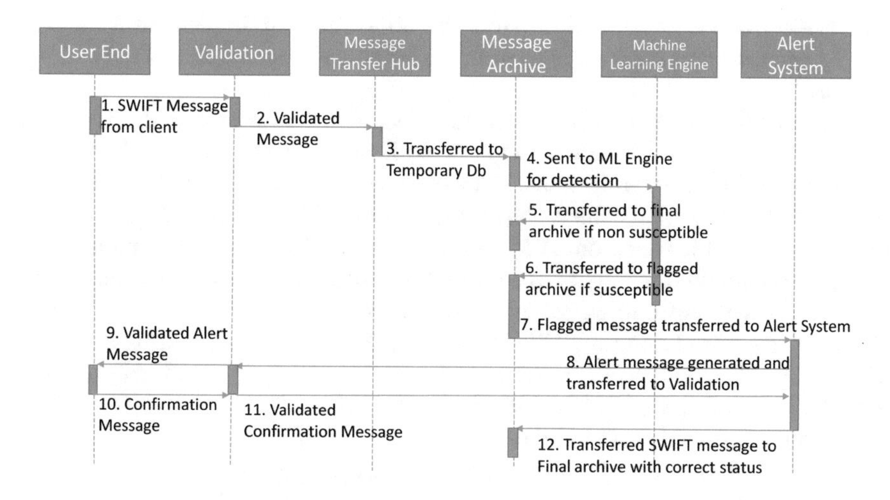

Fig. 7. Operational model

5.3 Security Analysis

We now discuss the security aspect of the system. SecureSWIFT is mainly proposed to use as a helper mode of communication along with the SWIFT network (Fig. 8).

Authentication: Each communication message from a user is signed digitally with the private key of that user so that the central system verify that using the public key, which is kept the user profile. This ensures the authentication of the sending of sending party.

Confidentially: In cases of sending messages, the message is encrypted

```
{1:F01MIDLGB22AXXX0548034693}{2:I1(
:20:8861198-0706
:23B:CRED
:32A:000612USD5443,99
:33B:USD5443,99
:50K:GIAN ANGELO IMPORTS
NAPLES
:52A:BCITITMM500
:53A:BCITUS33
:54A:IRVTUS3N
:57A:BNPAFRPPGRE
:59:/20041010050500001M02606
KILLY S.A.
GRENOBLE
:70:/RFB/INVOICE 559661
:71A:SHA
-}
```

Fig. 8. SWIFT message

with the receiver's public key. So, only the receiver can decode it using its own private key. Thus confidentiality is the preserver.

Integrity: As the sender digitally signs each message, the integrity is already addressed by the framework.

Non Repudiation: The system is built on asymmetric cryptographic techniques. As we know, asymmetric or public key cryptographic already ensures Nonrepudiation.

5.4 Experimental Setup and Results

Our system can be divided into three broad section for implementation. They are: **1)** Security and Messaging System, **2)** Machine Learning System, **3)** Archive System.

We consider following 6 fields as our input feature for the predictive model: ***Field 20,*** Sender's Reference, ***Field 23B,*** Bank Operation Code, ***Field 32A,*** Value Date/Currency/Interbank Settled Amount, ***Field 50,*** Ordering Customer, ***Field 59,*** Beneficiary Customer, ***Field 71A,*** Details of Charges. Our predictive model is based on SVM (Support Vector Machine). The model trains on these six mandatory fields. SVMs are helpful in text categorization. As real SWIFT messages of a financial institution is confidential and difficult to gather those, we generated dummy SWIFT messages for training and testing our model. We generated 1000 SWIFT messages of type "MT103". We trained our model with 80% of the data and tested the model with 20% of the data. We also conducted a ten fold cross-validation.

We evaluated our system in terms of accuracy. For accuracy calculation, we used Precision, Recall, and F1-score. True Positives are correctly classified transactions in the test set. That is, if the transaction is labeled as regular and our model classifies it as regular, then this transaction is a True Positive. False Positives are those transactions that are wrongly classified by the model. For example, if the model classified one transaction as irregular, but that transaction is actually regular in the test set, then the transaction will be considered as false positives. Our results (Table 4) showed that the proposed model achieved 67.4% precision and 73.9% recall on a average. The highest precision and recall we achieved were 71.4% and 77.9% for Fold-8.

Table 4. Results - precision & recall

	Precision	Recall	F1-score
Fold-1	67.8%	72.1%	69.9%
Fold-2	70.3%	74.2%	72.2%
Fold-3	69.8%	75.5%	72.5%
Fold-4	64.2%	76.2%	69.7%
Fold-5	65.9%	69.8%	67.8%
Fold-6	67.6%	73.4%	70.4%
Fold-7	66.1%	72.1%	69.0%
Fold-8	71.4%	77.9%	74.5%
Fold-9	67.5%	74.3%	70.7%
Fold-10	63.8%	70.6%	67.03%
Avg.	67.4%	73.9%	70.5%

6 Related Work

To our best of knowledge, this is the first survey on a digital bank heist. In literature, there are several studies done on the security of Internet/Online banking.

Lee et al. [31] presented a study about the security of Internet banking and private financial information in South Korea. Mannan et al. [33] conducted a survey among users to discover the gap in real-world online banking. The author in [43] examined the function and operation flow of the electronic fund transfer process as well as its security control mechanism. Also, some important security features of SWIFT were investigated in this research. Ahmed et al.[25] discussed the intrusion of the banking system and provided a recommendation to prevent intrusion. In [24], different types of online banking systems attacks were observed as well as the vulnerabilities of the systems were discussed. Claessens et al. [28] discussed the security of electronics, especially Internet and mobile banking systems, and presented an overview and evaluation for the techniques that were used in the systems. Three different kinds of case studies conducted in [24] to understand E-Banking phishing techniques and attack strategies. Yousafzai et al. [41] explored the trust-related issues and risk concerns among customers about electronic banking.

Apart from bank hacking, other online criminal activities were also explored. Motoyama et al. [34] empirically characterized six different underground forum. Chachra et al. [27] presented a measurement-based characterization of cookie-stuffing fraud in online affiliate marketing. An empirical study on keylogger-based stealing of credentials via dropzone was presented in [30]. Li et al. [32] proposed a deep learning-based framework for identifying top malware/carding sellers in underground forums. Pearce et al. [36] conducted a study to characterize large-scale click fraud in ZeroAccess, which is one of the largest click fraud botnets in operation.

As most of these attacks were launched using malware, we also did a literature review of financial malware. Samaneh et al. [39] explored the incentives and strategies of attackers by analyzing the instructions sent to machines infected with Zeus malware. Reicardi et al. [37] presented a system to map financial botnet networks and to provide a non-deterministic score to its associated elements. In [38], the author analyzed seven years of data on the command and control (C&C) infrastructure servers of botnets that have engaged in attacks on financial services. Oro et al. [35] proposed the use of publicly available IP blacklists to detect drones and command & control nodes that are parts of financial botnets. Alazab et al. [26] investigated the different techniques adopted by obfuscated malware as they are growingly widespread and increasingly sophisticated with zero-day exploits. Zhou et al. [42] characterized android malware from various aspects, including their installation methods, activation mechanisms as well as the nature of carried malicious payloads.

7 Conclusion

In the age of the Internet, the nature of cyber crime is evolving very quickly. In this paper, we looked into the emerging problem of digital bank heists to identify common patterns and vulnerabilities. Our study revealed several vulnerabilities in bank messaging systems such as SWIFT. We presented an in depth security analysis of several heists and discussed the deficiencies of the security system of financial institutions. Based on our observation, we proposed a cloud based service which can be used in detection of suspicious SWIFT messages and alert clients in time. We presented a detailed architecture and a complete operation model of the system. We developed a prototype using our proposed architecture. Our machine learning based predictive model achieved 67.4% precision and 73.9% recall. These accuracy could be improved by selecting better features from SWIFT messages. In future, we would like to include other features to achieve better accuracy.

Acknowledgements. This research was supported by the National Science Foundation through awards DGE-1723768, ACI-1642078, and CNS-1351038.

References

1. Ach Volume Grows. https://www.nacha.org/news/ach-volume-grows-56-percent-adding-13-billion-payments-2015-0
2. Ach vs Wire vs Electronic Transfer? What is the difference?. https://moneyrep.iaacu.org/2014/05/ach-vs-wire-vs-electronic-transfer-what-is-the-difference/
3. Citibank Credit Card Data Breach. https://www.thedailybeast.com/citi-credit-card-leak-nearly-twice-as-big
4. Clearing House Interbank Payments System. https://en.wikipedia.org/wiki/Clearing_House_Interbank_Payments_System
5. Cyber Security Cost of JPMorgan Chase. http://www.bankrate.com/finance/banking/us-data-breaches-1.aspx#slide=5
6. Fedwire Funds Services. https://www.federalreserve.gov/paymentsystems/fedfunds_about.htm
7. Hacker Bugging the System of Bangladesh Bank. https://web.archive.org/web/20160312145208/http://www.asianews.network/content/hackers-bugged-bangladesh-bank-system-jan-11271
8. How a Simple Typo Helped Stop a 1 Billion Dollar Digital Bank Heist - The Washington Post. https://goo.gl/Fm5NRa
9. Paypal. https://en.wikipedia.org/wiki/PayPal
10. RCBC had to Pay 1 Billion Filipino Dollar as Penalty. https://manilastandard.net/business/213132/rcbc-pays-half-of-p1-b-penalty.html
11. Rupay Credit Card Data Breach. http://www.reuters.com/article/us-india-banks-fraud/security-breach-feared-in-up-to-3-25-million-indian-debit-cards-idUSKCN12K0CC?il=0
12. Bangladesh Bank Heist (2015). https://en.wikipedia.org/wiki/Bangladesh_Bank_heist
13. Brazilian Bank Hack (2015). https://www.wired.com/2017/04/hackers-hijacked-banks-entire-online-operation/
14. Ecuador Bank Heist (2015). http://www.reuters.com/article/us-wells-fargo-banco-del-austro-ruling-idUSKCN12J03J
15. Fin-What is Swift (2015). https://fin.plaid.com/articles/what-is-swift
16. Hetachi Heist (2015). https://timesofindia.indiatimes.com/business/india-business/hitachi-payment-accepts-malware-hit/articleshow/57071322.cms
17. JP Morgan Bank Heist (2015). http://www.businessinsider.com/jpmorgan-hacked-bank-breach-2015-11

18. Lyod Bank Hack (2015). http://www.theinquirer.net/inquirer/news/3003091/lloyds-bank-hack-ddos-attack-disrupts-banks-online-services
19. Russian Bank DDoS Attack (2015). https://www.theregister.co.uk/2016/11/11/russian_banks_ddos/
20. Tesco Bank Heist (2015). http://thehackernews.com/2016/11/tesco-bank-hack.html
21. Ukraine Bank Swift Hack (2015). http://thehackernews.com/2016/06/ukrainian-bank-swift-hack.html
22. US Bank DDoS Attack (2015). http://www.nytimes.com/2012/10/01/business/cyberattacks-on-6-american-banks-frustrate-customers.html
23. Vietnam Bank Swift Hack (2015). http://www.reuters.com/article/us-vietnam-cybercrime-idUSKCN0Y60EN
24. Aburrous, M., Hossain, M.A., Dahal, K., Thabtah, F.: Experimental case studies for investigating e-banking phishing techniques and attack strategies. Cogn. Comput. **2**(3), 242–253 (2010)
25. Ahmad, M.K.A., Rosalim, R.V., Beng, L.Y., Fun, T.S.: Security issues on banking systems. Int. J. Comput. Sci. Inf. Technol. **1**(4), 268–272 (2010)
26. Alazab, M., Venkatraman, S., Watters, P., Alazab, M., Alazab, A.: Cybercrime: the case of obfuscated malware. In: Global Security, Safety and Sustainability & e-Democracy, pp. 204–211. Springer (2012)
27. Chachra, N., Savage, S., Voelker, G.M.: Affiliate crookies: characterizing affiliate marketing abuse. In: Proceedings of the 2015 ACM Conference on Internet Measurement Conference, pp. 41–47. ACM (2015)
28. Claessens, J., Dem, V., De Cock, D., Preneel, B., Vandewalle, J.: On the security of today's online banking systems. Comput. Secur. **21**(3), 253–265 (2002)
29. FBI: Bank Crime Statistics (2015). https://www.fbi.gov/investigate/violent-crime/bank-robbery/bank-crime-reports
30. Holz, T., Engelberth, M., Freiling, F.: Learning more about the underground economy: a case-study of keyloggers and dropzones. Comput. Secur.-ESORICS **2009**, 1–18 (2009)
31. Lee, J.H., Lim, W.G., Lim, J.I.: A study of the security of Internet banking and financial private information in South Korea. Math. Comput. Model. **58**(1), 117–131 (2013)
32. Li, W., Chen, H.: Identifying top sellers in underground economy using deep learning-based sentiment analysis. In: 2014 IEEE Joint Intelligence and Security Informatics Conference (JISIC), pp. 64–67. IEEE (2014)
33. Mannan, M., van Oorschot, P.C.: Security and usability: the gap in real-world online banking. In: Proceedings of the 2007 Workshop on New Security Paradigms, pp. 1–14. ACM (2008)
34. Motoyama, M., McCoy, D., Levchenko, K., Savage, S., Voelker, G.M.: An analysis of underground forums. In: Proceedings of the 2011 ACM SIGCOMM Conference on Internet Measurement Conference, pp. 71–80. ACM (2011)
35. Oro, D., Luna, J., Felguera, T., Vilanova, M., Serna, J.: Benchmarking IP blacklists for financial botnet detection. In: 2010 Sixth International Conference on Information Assurance and Security (IAS), pp. 62–67. IEEE (2010)
36. Pearce, P., Dave, V., Grier, C., Levchenko, K., Guha, S., McCoy, D., Paxson, V., Savage, S., Voelker, G.M.: Characterizing large-scale click fraud in ZeroAccess. In: Proceedings of the 2014 ACM SIGSAC Conference on Computer and Communications Security, pp. 141–152. ACM (2014)

37. Riccardi, M., Oro, D., Luna, J., Cremonini, M., Vilanova, M.: A framework for financial botnet analysis. In: ECrime Researchers Summit (ECrime), 2010, pp. 1–7. IEEE (2010)
38. Tajalizadehkhoob, S., Gañán, C., Noroozian, A., Eeten, M.V.: The role of hosting providers in fighting command and control infrastructure of financial malware. In: Proceedings of the 2017 ACM on Asia Conference on Computer and Communications Security, pp. 575–586. ACM (2017)
39. Tajalizadehkhoob, S., Asghari, H., Gañán, C., van Eeten, M.: Why them? Extracting intelligence about target selection from Zeus financial malware. In: WEIS (2014)
40. Financial Crimes Enforcement Network of Department of the Treasury (2015). https://www.fincen.gov/reports/sar-stats
41. Yousafzai, S.Y., Pallister, J.G., Foxall, G.R.: A proposed model of e-trust for electronic banking. Technovation **23**(11), 847–860 (2003)
42. Zhou, Y., Jiang, X.: Dissecting android malware: characterization and evolution. In: 2012 IEEE Symposium on Security and Privacy (SP), pp. 95–109. IEEE (2012)
43. Zhu, D.: Security control in inter-bank fund transfer. J. Electron. Commer. Res. **3**(1), 15–22 (2002)

Identifying Vulnerabilities in Security and Privacy of Smart Home Devices

Chola Chhetri(✉) and Vivian Motti

George Mason University, Fairfax, VA 22030, USA
{cchhetri,vmotti}@gmu.edu

Abstract. Smart Home Devices (SHDs) offer convenience that comes at the cost of security and privacy. SHDs can be subject to attacks and they can be used to conduct attacks on businesses or governments providing services to individuals. In this paper, we report vulnerabilities that have been published in research papers in IEEE Xplore digital library and ACM digital library. We followed a systematic approach to search for vulnerabilities in the literature, analyzed them and placed them in common categories. The study resulted in 153 vulnerabilities. The categories are based on the place of occurrence or component of smart home architecture, such as device, protocol, gateway, network, and software architecture. We also identified areas of research and development that have been underexplored in the past and need further efforts. Researchers, developers and users will benefit from this comprehensive analysis and systematic categorization of smart home vulnerabilities.

Keywords: Vulnerabilities · Smart Home Devices · Internet of Things · Security · Privacy

1 Introduction

With over 20 billion Internet of Things (IoT) devices projected to be used globally by 2020, home automation is changing the way people interact and live, through the use of intelligent voice assistants, smart door locks, light bulbs, etc. [32]. Smart Home Devices (SHDs) pose threats to the security and privacy of individuals, businesses, and the society [5,13,72]. Unauthorized actors have hacked baby monitors [72], IoT search engines have provided public access to baby videos [29], toys have leaked parent-child conversations [38], drones have controlled home lights by flying above houses [13], and malware attacks on vulnerable IoT devices have brought down prominent Domain Name System (DNS) infrastructure affecting many large businesses, such as Dyn [5]. Information about smart home vulnerabilities (SHV) is scattered across a large number of research articles in databases. This paper systematically catalogs SHVs from research articles published in IEEE digital library and ACM digital library from 2010 to 2019.

K.-K. R. Choo et al. (Eds.): NCS 2020, AISC 1271, pp. 211–231, 2021.
https://doi.org/10.1007/978-3-030-58703-1_13

Security vulnerabilities are closely linked with privacy, as they can lead to privacy violations. For instance, a security vulnerability may allow an adversary to hack into a baby monitor, and get audio, as well as video recordings of a child, thus violating the child's privacy. Hence, addressing SHD vulnerabilities helps resolve not only security concerns but also privacy concerns. Furthermore, unaddressed concerns leads to non-adoption and rejection of technology, such as the past case in Netherlands, where a rollout of smart meters had led to failure [30].

To the best of our knowledge, this paper is the first to systematize SHV information through literature review. Anwar et al. (2017) proposed a taxonomy for smart home that broadly classified smart home security threats into three types: (a) intentional/abuse, (b) malfunctions/failures, and (c) unintentional. Intentional threats include threats from an adversary, e.g. denial of service, identity fraud, manipulation of information, eavesdropping/traffic hijacking. Malfunctions include interruptions or disruptions caused by failures in devices, communication, network, power, third party services, and Internet. Unintentional threats are accidental abuses, such as accidental sharing of sensitive data, policy flaws, design flaws, among others [6].

Anwar's taxonomy serves as a preliminary frame for reference. However, its scope was limited to a small set of the literature (15 references) and the threats classified were non-device-specific [6]. Mosenia and Jha (2016) studied 20 IoT security threats and divided them into three layers: Edge nodes (Computing nodes and radio frequency identification), Communication, and Edge computing [55].

Our comprehensive study of SHD vulnerabilities synthesizes 153 SHVs, categorizes them, the scope of attacks in the smart home, and identifies research areas that need further exploration. Our study is a part of research aimed at exploring SHVs and designing user-centric privacy controls for SHDs [15]. In this paper, we report on the following:

1. We perform a systematic literature review of **119** articles and catalog SHV information published in selected databases (Sect. 3).
2. We categorize the SHVs into four categories, each category containing subcategories. This could serve as vulnerabilities taxonomy for the smart home domain (Sect. 3).
3. We synthesize the solutions to SHVs from our literature review (Sect. 4). Some solutions are specific to a vulnerability and/or a device, while others are general and do not address a specific vulnerability.
4. We identify research gaps and opportunities in the security and privacy of smart home (Sect. 5).

2 Methodology

In this section, we describe the scope, inclusion criteria and exclusion criteria of our systematic literature review.

2.1 Databases and Keywords

We performed full-text literature search on IEEE Xplore Digital Library[1] and ACM Digital Library[2] from October 2018 to March 2019. We used the following keywords in the search:

- smart AND home AND vulnerabilities
  - In Abstract
  - In Document Title.

2.2 Inclusion and Exclusion Criteria

We read all papers in the search results to determine if they were relevant to the study. For a paper to be considered as 'relevant', it had to discuss or contain information about smart home vulnerabilities (SHV). We read the abstract first and then the paper to determine SHV content. If a paper did not contain SHV information, we considered it 'irrelevant' to the study.

For each paper in the search results, the following steps were followed in order:

1. Read abstract.
2. If the abstract contains SHV information, include it.
3. If the abstract does not contain SHV information, read the paper.
4. If paper contains SHV information, include it.
5. If paper does not contain SHV information, exclude it.

Outcomes: The search in the two databases returned a total of 119 papers, among which 98 papers were included and 21 were excluded (see Table 1). Publication dates ranged from 2010 to 2019. ACM papers were published in 33 proceedings and one periodical. IEEE papers were published in 73 conferences, and 11 journals and magazines.

Table 1. Number of papers included in and excluded from the study.

Source	# Papers	# Included	# Excluded
IEEE	85	73	12
ACM	34	25	9
Total	119	98	21

For papers included in the study, a systematic cataloging of SHV information was done, following a template we developed for this purpose. In the catalog, we included the name of the vulnerability, name/type of device, explanation of the vulnerability, solution, explanation of the solution, who implements the solution (user or developer), and drawbacks of the solution, when available in the paper.

[1] ieeexplore.ieee.org.
[2] dl.acm.org.

3 Vulnerabilities of the Smart Home

In this section, we discuss the vulnerabilities found in the study. The cataloging process resulted in 153 SHVs. SHVs existed in various parts of the SH network, including the device, SHD applications (such as voice assistants), software architectures and frameworks, communication protocols (such as WiFi, 802.15.4, Zigbee, Routing for Low Power and Lossy Network (RPL), Precision timing protocol (PTP), etc.), smart home network, operating system (such as Android), and authentication systems (such as Zigbee Light Link).

Among the SHVs, 75 were device specific. The papers identified the devices with the vulnerability discussed. Devices with vulnerabilities included cameras (such as TP Link), Belkin motion sensor, Withings scale, light bulbs (Philips Hue, LIFX), Chromecast, Google home, Hello Barbie talking doll, Haier Smartcare, HP Envy printer, hubs/controllers, TP Link power switch, thermostat (Nest), smart meters, smart speakers, SmartThings, and Voice Control System. The 78 remaining vulnerabilities were generic (non-device-specific) and related to communication protocols, software architectures, communication, voice assistants (such as Alexa), operating systems, applications, and authentication mechanisms.

3.1 Device Vulnerabilities

The papers included in our study revealed vulnerabilities in SHD hardware or in the software running in the device. We divided the device vulnerabilities in eight categories:

1. Authentication Vulnerabilities
2. Information Leakage or Disclosure
3. Data Protection Vulnerabilities
4. Data Manipulation
5. Voice Interface Vulnerabilities
6. User Behavior Detection
7. Service Disruption
8. Other Vulnerabilities.

Authentication Vulnerabilities. Authentication in an SHD ensures that only legitimate user or software process have access to the device features, control and operation [20]. Authentication vulnerabilities include lack of authentication, default credentials, hard-coded credentials, leaked credentials, weak authentication, flawed authentication protocol, and de-authentication attack. We will describe each of these briefly in the following sub-sections:

Lack of Authentication. Ma et al. (2018) showed that an attacker can post a message (text, audio, video) on the user's TV/Chromecast screen without requiring any authentication [48]. Mahadewa et al. (2018) demonstrated that any SHD in a home network can control the light bulbs available. Chromecast allowed private YouTube videos to be cast on television without requiring any authentication [49].

Default Credentials. While the SHD industry as a whole has not caught up in implementing authentication credentials in devices [48,49], some manufacturers allow the set up of credentials, such as passwords and pins, in their devices. However, many devices run with their default credentials [31,54] and users are not aware about how to change them. Adversaries can easily find default credentials on the Internet. There are also search engines available that allow public access to SHDs (such as cameras) online [12]. Papers included in our study reported the use of default credentials in cameras [4,60], routers [51], thermostats [54], plugs [43], printers [63], light bulbs and motion sensors [63]. An adversary does not need to be technically savvy to learn default credentials. S/he needs to be able to search the Internet and have access to an Internet-connected device.

Hard-Coded Credentials. When credentials are hard-coded into the device and changes are not allowed, the device becomes vulnerable to attacks irrespective of other security and privacy mechanisms implemented in the device [12,43]. SHDs with image, audio and video recording capability can easily compromise the privacy and security of individuals, if they are operate with hard-coded credentials that can not be changed by the user.

Leaked Credentials. Saleh et al. (2018) demonstrated that credentials (username and password) of motion sensor and a Closed-circuit television (CCTV) camera were leaked to any observer of smart home traffic. Authentication credentials were not protected from being visible to a passive observer [60].

Weak Authentication. Prior research presents evidence of poorly implemented authentication, that allows an adversary to gain access to cameras [41,60,63]. Saleh et al. (2018) conducted penetration testing on a camera used to monitor a home smart meter using the Kali[3] operating system, which allowed the investigators access to the camera using default username-password combination [60]. This means an adversary could have full access to the smart camera without the user's knowledge. Authors claim that surveillance cameras of this type are used to monitor smart grid for security. Vulnerable CCTV cameras, thus lead to vulnerable smart grids [60].

According to Lei et al. (2018), Alexa used a voice word (analogous to a password) for user authentication. However, the person speaking the voice word did not have to be an authenticated user; it could be any one who knew the voice word [41]. Similarly, Sivanathan et al. (2017) showed that an attacker was able to send commands to control a light bulb, a power switch and a printer due to poor authentication [63]. In another experiment, Alharbi and Aspinall (2018) found that the Web interface of a camera used a weak password policy, did not require complex passwords and was prone to a brute-force password attack [4].

Flawed Authentication Protocol. In some SHDs, such as Philips hub, authentication is implemented; however, the authentication protocol used has security

[3] kali.org.

flaws. Mahadewa et al. (2018) found that the Philips Hue hub generates authentication tokens for all devices, whether they are authenticated or not. This results in unauthenticated and malicious devices being able to connect to Philips hub and control the smart home network or eavesdrop [49].

De-authentication Attack. Sun et al. (2018) showed that sending 802.11 de-authentication frame to disconnect a light bulb from the access point disabled home Internet connection or forced light bulb to connect to rogue access point. The light bulb was designed to remember the state (on/off) in case it was disconnected, and it lacked a physical power switch (on/off). Consequently, if the attacker turned off the light bulb and then performed de-authentication attack, the user could not turn it back on. In the same experiment, a camera was also found vulnerable to such de-authentication attack, which could render the camera unusable to the user [65].

Information Leakage or Disclosure. Papers included in our analysis show the potential of leakage of information collected by SHDs [4,36,65,67]. We have divided the information leakage or disclosure-related vulnerabilities into the following 5 sub-categories:

Log File Information Leakage. Alharbi and Aspinall (2018) found that the Android app of a camera stored the personal information of end users (such as home address, encryption keys, and WiFi credentials) in a log file, easily accessible to an adversary [4]. They also showed that in some SHD apps, system crashes could lead to leakage of sensitive data, such as device Unique Identifier (UID), email address, phone number, Global Positioning System (GPS) location, text messages, and log messages [4]. Johnson et al. (2018) claimed that SHD vendors could write this information from Android log file to another file and malicious apps could deliberately cause crashes to obtain this information. This information could be used for user tracking, user behavior prediction, and location determination [36].

Device Information Leakage. SHDs have unique identifiers, such as Media Access Control (MAC) addresses and serial numbers. When revealed, such identifiers can be used to permanently track the device and/or its owner. Alharbi and Aspinall (2018) found that cameras were revealing MAC addresses, device serial numbers, and device passwords, making it convenient for an adversary to access the camera without any sophisticated attack [4].

Personal Information Leakage. Tekeogl and Tosun (2015) showed that an adversary could passively listen to a Chromecast device and obtain unencrypted information, which included google account username, video id, time, operating system (OS) name, OS version, device brand and model. In addition, they conducted black box tests to reveal the leakage of remote information (name, brand, model, OS name, and OS version) to an observer [68].

Information Disclosure. Sivanathan et al. (2017) showed that unencrypted messages (including audio and video) were disclosed by smart home devices, such as Belkin Motion Sensors, TP Link cameras, Withings scales, and Phillips Hue light bulbs [63]. They also found that an Internet Protocol (IP) printer exposed the last scanned document to an attacker who controlled the printer via its web interface [63]. Bugeja et al. (2018) found similar information disclosure vulnerabilities in 'connected' cameras [12].

Device and Occupant Localization. Vulnerable smart cameras connected to the home network can reveal sensitive information not just about location but also about occupants in the home. Sun et al. (2018) showed that an adversary could perform traffic analysis by passively sniffing home network traffic to obtain knowledge about the number of occupants and the in-house location of occupants. The attacker would also not risk being detected due to the attack's passive nature [65]. Jia et al. (2017) showed that an adversary could effectively perform geo-location prediction in Google Home [34].

Data Protection Vulnerabilities. It is important that data in a smart home be protected. The data life cycle in a smart home includes collection from SHDs, transmission to hub and/or cloud, storage in hub and/or cloud, and processing [16]. In our categorization, data protection vulnerabilities include Lack of Encryption, Weak Encryption, and Weak Server-side Protection.

Lack of Encryption. True end-to-end encryption, when implemented well, can provide confidentiality of data in transit and also at rest in a storage device or cloud. Most smart phones today have feature of encryption; however, this feature may not be enabled by default [17]. Zhang et al.(2016) mentioned that most phones lacked encryption, which allowed an adversary to obtain SHD data easily in case the adversary established physical or remote access to the phone controlling SHDs [73]. Alharbi and Aspinall (2018) found that a Ring doorbell smart camera lacked encryption of video stream from the camera to server [4].

Weak Encryption. Encryption is a commonly discussed solution for sensitive data protection [8]. However, weakly implemented encryption can provide an additional attack vector in case an adversary were to break the encryption. For instance, Sivanathan et al. (2017) showed that a TP Link power switch could be easily broken into due to weak encryption [63]. Encryption vulnerabilities found in the literature review also included plain text key exchange in camera apps [4] and clear text communication from device to cloud in a Chromecast device [68].

Weak Server-Side Protection. Attackers could easily steal personal information and listen to conversations in a Hello Barbie talking doll due to the lack of data protection in the server [63]. Server-side data protection is an important issue. Users of SHDs tend to trust their vendors in protecting their data [16]. In the event that data related to users is breached, companies not only lose trust of their customers but also bear huge financial losses [1].

Data Manipulation. This category of SHD vulnerabilities includes vulnerabilities that allow an adversary to change configurations, alter data or modify applications in a home network. Pricing cyberattack falls under this category.

Pricing Cyberattack Past research has demonstrated that an adversary could change the data regarding electricity usage by gaining access to a smart meter [34,47]. This could result in altered utility bill (e.g. higher electricity bills) for a smart meter user and an attacker could reduce his/her bill but increase community peak energy usage as well as other users' energy bill [34]. Various detection frameworks have been proposed to solve this problem. Researchers have evaluated vulnerabilities in some presented frameworks, such as the lack of net metering in detection frameworks [47].

Voice Interface Vulnerabilities. The domain of smart speakers and voice enabled devices has presented new vulnerabilities in the SHD domain. This category includes vulnerabilities in devices with Voice User Interface (VUI).

Voice Command Attack. According to Alanwar et al. (2017), audible voice commands (generated by devices such as televisions) and inaudible voice commands (generated by malicious speakers) were able to activate smart speakers and force them to perform actions even when the legitimate user was not present [3]. An adversary with access to speaker-enabled devices in a smart home network could issue malicious commands, leading to fake transactions, burglary, or other unintended actions. For example, in 2015, Amazon Echo devices in users' homes played Christmas music in response to a television advertisement for Alexa, which confused and frustrated many users [11].

Hidden Command Attack. Voice commands can be hidden in a way that they appear as noise to human ears. Meng et al. (2018) found that an attacker was able to conduct a spoofing attack on a Voice Command System (VCS) and issue a hidden voice command to an SHD [52]. Hence, what appears as noise could be a command given by an adversary to open a garage door, or to unlock the house, or turn up the thermostat [52].

Inaudible Command Attack. Meng et al. (2018) also found that voice commands can be made inaudible to the human ears, and a spoofing attack in which an attacker inputs inaudible voice commands to a VCS is known as inaudible command attack [52].

Replay Audio (RA) Attack. Replay audio attack is a spoofing attack in which the attacker uses pre-recorded voice of a user to fool the voice control system of a SHD [52]. Malik et al. (2019) found that smart speakers such as Amazon Echo and Google Home were subject to replay attacks, in which an adversary could place fake orders, reveal personal information (such as the owner's name), and control IoT devices, such as smart doors [50].

User Behavior Detection. Apthorpe et al. (2017) showed that an observing entity, such as a service provider or an adversary, can analyze traffic generated from SHDs to predict the user presence (i.e. whether a user is home), sleep patterns, appliance usage patterns, occupancy patterns (i.e. how frequently a user is home), and the frequency of user motion [9]. In this paper, we refer to this inference as user behavior detection.

Pattern-of-Life Modeling. Beyer et al. (2018) set up a test bed network including a camera, outlet, motion sensor, and television, and used pattern-of-life analysis tool to analyze data leakage. They found that an adversary could infer the types of SHDs used in the home, identify events (such as user presence), track the user, map the smart home network, and gain access to the home [10].

User Presence Detection. Gong and Li (2011) showed that smart meters with differential transmission scheme (DTS) were vulnerable to user presence detection and that an adversary eavesdropping the traffic could infer whether the user is home [27]. DTS is a method for tracking electricity usage of a consumer by reporting power consumption to the utility company only when consumption changes. Since transmission frequency is proportional to power consumption, the attacker can reveal user presence by observing this transmission [27]. Li et al. (2012) discussed a similar attack on smart meters and called it Presence Privacy Attack (PPA) [42].

Service Disruption. Any vulnerability on SHD that can cause partial or complete interruption of access to SHD or its service is categorized as Service Disruption. Hence, this category includes jamming, denial of service, impersonation and replay.

Jamming. An adversary can disrupt network communication by introducing powerful jamming signals leading to interruption of service and battery drains [39]. Jamming attacks were previously demonstrated using smart meters [56].

Denial of Service. Denial of Service (DoS) attack on 802.15.4 Media Access Control (MAC) in Zigbee devices was shown to cause disruption of service [70].

Impersonation. Smart meters were shown to be vulnerable to impersonation, where an adversary could introduce rogue (or fake) device to appear legitimate [56].

Replay. In LIFX lightbulb system, an attacker was able to intercept and replay User Datagram Protocol (UDP) packets to eavesdrop the network and control a light bulb [49]. Feng et al. (2017) showed that an attacker could replay pre-recorded video frames without motion to replace those with motion, to compromise the alert/alarm system of security cameras and hide the malicious behavior [22].

Other Vulnerabilities. Twenty-six other vulnerabilities were found in various SHDs. In Table 2, we show the names of these additional vulnerabilities, the devices they were found in, and the references to related articles.

Table 2. Vulnerabilities in SHDs.

Vulnerability	Device	Reference
Access to Remote Data	Nest Thermostat	[54]
Account Lockout Mechanism	Camera	[4]
Authorization Code Compromise	Google Home	[35]
Brute-force attack	Plug (Edimax SP-2101W)	[43]
Cross-protocol vulnerability	Cross-protocol devices	[63]
Cross-site request forgery	Camera	[12]
Cross-site scripting	Camera	[12]
Design flaw	Haier SmartCare	[71]
Device scanning attack	Plug (Edimax SP-2101W)	[43]
Firmware attack	Plug (Edimax SP-2101W)	[43]
Flawed/lacking TLS implementation	Camera	[4]
Home Area Network Id (HANID) conflict attack	Smart meter	[24]
Intrusion	Smart grid	[53]
Key management (SILDA protocol)	Smart grid	[62]
Lack of Control to Administration Commands	Hub	[49]
Light bulb attack	Lightbulb	[69]
Mis-response to Discovery Request	Hub, Chromecast	[49]
Overprivilege	SmartThings	[28]
Over-privileged app	Camera	[4]
Rogue controller injection	Z-wave controller/gateway	[25]
SEP vulnerabilities	Smart grid	[7]
SH network compromise	Google Home	[35]
Spoofing attack	Plug (Edimax SP-2101W)	[43]
Unnecessary open ports	Camera	[4]
Unprotected WiFi Hotspot in SHD	Lightbulb	[49]
Use of Insecure Underlying Protocols	Hub	[49]

3.2 Application Vulnerabilities

SHDs are usually controlled by users through associated applications running on smartphones. We found two application vulnerabilities in our literature review: home network infiltration and data leakage.

Home Network Infiltration. Malicious apps can make their way through app stores and then to the home network to cause larger attacks such as distributed denial of service (DDoS) [64].

Data Leakage. Ahmad et al. (2015) showed that mobile OSs, such as Android, had limitations that allowed data leakage from end-user devices to vendor servers as well as unintended recipients [2].

3.3 Communication Vulnerabilities

In this section, we discuss Data Leakage and Protocol Vulnerabilities.

Traffic Analysis. Sanchez et al. (2014) showed that even with proper encryption, an attacker could analyze WiFi traffic to infer sensitive information, such as inventory of SH devices, device functions and relationships, and user behavior [61].

Protocol Vulnerabilities. Past research has revealed weaknesses in the RPL protocol [26] and proposed improvements [18,26]. Fan et al. (2019) discussed time synchronization issues in PTP protocol (also called IEEE 1588 Precision clock synchronization protocol), which could cause inaccurate device functions due to device receiving wrong time information [21].

Past research also shows the possibility of following attacks on SHD communication:

- Jamming: An adversary can disrupt the network communication by introducing strong jamming signals leading to denial of service attack as well as SHD battery drains [39].
- Guaranteed Time Slot (GTS) attack: An attacker can disrupt the communication between SHD and its gateway, causing collision, corruption and retransmission of packets. This leads to a loss of communication and DoS attack [39].
- Acknowledgement (ACK) attack: An attacker eavesdrops communication, hijacks packet and sends fake ACK to trick the sender. This leads to the attacker taking control over the smart home network communication [39].
- XMPPloit: XMPPloit can force a SHD to not encrypt the communication, allowing eavesdropping and data leakage [39].
- Eavesdropping: Unencrypted communication allows an attacker to decipher sniffed communication causing a breach of confidentiality [33].
- Denial of Service: Attacker can use malicious traffic to render the home network unresponsive and the user cannot access the home network services [33].

3.4 Software Architecture Vulnerabilities

Liu et al. (2017) found the following vulnerabilities in the Joylink home automation architecture [44]:

WiFi Credential Theft. WiFi credentials were transmitted after being encoded one character at a time following the IP address. This allowed an adversary to easily steal WiFi credentials and access home WiFi without authentication [44].

Vulnerable Crypto Key Management. The Joylink architecture utilized a vulnerable key generation technique and a local adversary could launch a man in the middle (MiTM) attack [44].

Traffic Decryption. The Joylink architecture utilized a vulnerable crypto key management technique and a local adversary could launch MiTM attack and decrypt all traffic, thus allowing breach of confidentiality of sensitive information [44].

Device Hijacking. The Joylink architecture utilized a weak communication security and a local adversary can obtain the MAC address of a user device, log into its own cloud account, hijack the device and control it remotely [44].

Out of Band Device Control. An attacker was able control the SHD by creating a fake server, without accessing the cloud account or the app [44].

Device Impersonation. The Joylink architecture's lack of authentication allowed an attacker to log in to a cloud account, activate a user device with a spoofed MAC address that was easy to obtain due to poor crypto management [44].

Firmware Modification. The Joylink architecture's lack of verification of downloaded firmware made it vulnerable to malicious firmware from attackers [44].

Visible Data/Communication. Control commands and uploaded data were visible to an observer, that could lead to private information being revealed [44].

WiFi Credentials on the Cloud. WiFi credentials were uploaded to the cloud server, even when there was no need for this private user information to be sent to the cloud [44].

Weak Key. The Joylink architecture used a timestamp value as an Advanced Encryption Standard (AES) key, made it easy for an adversary to predict the key (due to a small key space) and to reveal information collected by the SHD app [44].

Reverse engineering and source code analysis of SmartApps performed by Fernandes et al. (2016) showed that more than half of the 499 apps were **over-privileged**, and apps retained unnecessary permissions even when the user denied them [23]. OpenHAB[4] and IoTOne were presented as solutions to this issue [28].

Web attacks, such as **SQL injection** and **unauthorized access to sensitive data**, were possible due to poor use (or exploitation) of Application Programming Interfaces (API) in the Spring framework, an open source framework for apps.

4 Solutions to SHD Vulnerabilities

In this section, we discuss the solutions to SHD vulnerabilities proposed by investigators of past research.

Main solutions to **authentication vulnerabilities** in SHDs included the use of the following authentication mechanisms [41,43,49,60,60,63]:

1. Requiring the use of credentials, such as username-password combination,
2. Enforcing the change of credentials,
3. Protecting the authentication credentials, and
4. Ensuring that the authentication protocols used are up-to-date, strong and unbroken.

Two-factor authentication has been investigated as a potential approach to strong authentication in SHDs. Crossmand and Liu (2015) proposed the Smart Two-Factor Authentication, in which a company gives its user a smart card that produces (and stores) a token for two-factor authentication at the request of the user [19]. Researchers have recommended that SHDs must have a physical

[4] https://www.openhab.org/.

switch for the user to manually turn the device on/off, and a default fail-over state so that an adversary can not render the device unusable [65].

Solutions to **information leakage and exposure** included limiting or restricting the use of personal information for logging and debugging purposes [4], protecting sensitive device information [4], and encrypting sensitive information [4,25,73].

Data collected by SHDs needs to be protected in all stages: collection, transmission (device to hub, hub to cloud), storage (in hub or cloud), and processing. Encryption is often used to protect data, but it needs to be implemented without flaws and encryption keys need to be protected too [4,25,33,54,59,73]. Salami et al. (2016) proposed the Lightweight Encryption for Smart homes (LES), an encryption technique with low overhead and computation requirement, and identity-based stateful key management [59]. Further research is needed though to evaluate the use of such encryption techniques in various categories of SHDs.

Maintaining data integrity in smart meters is crucial for the accuracy of electricity bill and protection of smart meters and the smart grid from data manipulation attacks. Various pricing cyber attack detection frameworks have been proposed, such as the Electricity Pricing Manipulation Detection Algorithm [46], Partially observable Markov decision process (POMDP) based smart home pricing cyberattack detection framework [47], and single event detection technique based on support vector regression [45].

As a solution to audio vector attacks, Alanwar et al. (2017) proposed EchoSafe, a sound navigation ranging (SONAR) based active defense mechanism, that checks for the presence of the user as soon as the smart speaker is activated, to ensure that commands are executed only if the user is present nearby (in the room) [3]. To protect from hidden and inaudible command voice attacks, Meng et al. (2018) proposed Wivo, a tool that authenticates the voice input with mouth motions of the user (liveness detection) [52]. Replay attack detection approaches include Higher-order spectral analysis (HOSA)-based replay attack detection approach [50] and Wivo [52].

User presence attack and behavior inference attack are usually mitigated by introducing fake traffic into a user's smart home network to minimize the likelihood of an inference to occur. Beyer et al. (2018) proposed a technique called MIoTL (Mitigation of IoT Leakage), which introduces fake traffic to the home network using (a) a device shadow to protect from identifying and classifying devices, and (b) a MAC shadow to mitigate user tracking [10]. Gong and Li (2011) proposed a similar approach for addition of null packets to the smart meter traffic during idle times to emulate busy times, confuse the observer, and mitigate user presence detection [27]. Li et al. (2012) proposed a similar method called Artificial Spoofing Packet (ASP), which added dummy packets to the transmission to trick an eavesdropper and mitigated user presence detection [42].

To mitigate denial of service attack in the Zigbee protocol, Whitehurst et al. (2014) proposed integrity checks on received packets (including acknowledgments) to make it difficult for an adversary to forge packets [70]. Namboodiri et al. (2014) proposed SecureHAN to combat jamming, impersonation, replay, and repudiation attacks [56]. Feng et al. (2017) showed that video replay attack in cameras could be thwarted by hardware isolation of the motion detection module [22].

To prevent home network infiltration via SHD apps, developers could employ network traffic analysis [64]. Data leakage through apps could be prevented by restricting app downloads only from home automation app stores, and by establishing fine grained policies to improve the communication between home automation apps and non-home-automation apps [36].

Finally, 5 articles in our data set have presented intrusion detection systems (IDS) as a method of protecting the home network [14,24,57,58,66]. The proposed IDS solutions are proofs of concept and prototypes. SHD users will benefit from fully functional tools available for consumer use.

5 Discussion

The systematic literature review showed 153 vulnerabilities in 75 devices. We found little consensus in the naming of the vulnerabilities. So, we categorized them based on vulnerabilities characteristics and similarities of the attacks. Based on the architectural components these vulnerabilities were found in, the smart home network presents many attack surfaces, making it harder to fully protect the home network from adversaries. The attack surface of a home network includes:

- smart home device, including hardware, operating system, and applications,
- communication protocols, that run on the SHD, controller or hub, and home router or gateway,
- smart phones used to control SHDs, including phone hardware, operating system, and applications,
- home automation software or software framework, and
- software securing the home network.

The attack surface increases as the number of devices and features in the smart home grows. New methods of attack are also introduced, such as home burglary and fake orders through attacks on VUI [40]. Thus, providing security and privacy in the smart home is challenging.

It is clear that the authentication vulnerabilities category was the largest, with 8 sub-categories. The lack of authentication poses threats to a smart home. An attacker may be able to control someone's door lock, garage door opener, thermostat or coffee maker and issue malicious commands. Evidently, there is a need for SHD manufacturers to implement authentication properly to address this issue and allow only authorized users to control an SHD.

In Sect. 3.1, we presented that research literature showed default credentials vulnerability was found in 7 SHDs. It is argued that market competition and the pressure to create low-cost SHDs in a short time frame has led to the issues lacking or poorly implemented authentication, which provide an additional (and easy) vector for adversaries to enter into the private home network. Baby monitor hacks [72] and Mirai botnet [37] that appeared widely in news were primarily possible due to default credentials. In order to prevent large scale attacks like Mirai in the future, default credentials vulnerability must be addressed in all SHDs. The issue of lack of authentication can be addressed by manufacturers requiring authentication credentials in their SHDs, and by developers requiring the user to change default credentials during device setup [4,51].

Past user studies have shown that users care about data protection and their privacy, but trust the vendors to provide appropriate data security and privacy protection [16,38,72]. Manufacturers need to reduce vulnerabilities in their SHDs by following secure development practices and also use up-to-date, secure protocols for communication with other devices to reduce cross-protocol vulnerabilities.

It is challenging for SHDs with limited memory and computational power to locally encrypt data before sending to a cloud server. The data in transit, thus, risks potential breach of confidentiality. So, encryption techniques suited for such environments need to be evaluated.

Next, we will summarize open research areas in the area of smart home security and privacy, and discuss the limitations of our work.

5.1 Open Research Areas

We have identified the following open research areas in the security and privacy of SHDs:

- Security analyses of SHDs, protocols, and software frameworks, especially of newer models, to find out vulnerabilities and develop solutions.
- Authentication mechanisms suitable for low power, low-resource, low-cost SHDs.
- User-centric methods of enforcing change of default credentials and setting up strong credentials.
- Secure methodologies for storing and managing credentials in the smart home network.
- Data protection techniques, such as encryption, customized to SHD environment.
- Study of whether data manipulation attacks, similar to those in smart meters, are possible in other SHDs, and development of solutions, if necessary.
- Development and evaluation of solutions to voice interface vulnerabilities, as VUI attacks are evolving with the popularity of voice interfaces.
- Best practices in data protection at various stages of data life cycle including use, rest and transit.

5.2 Limitations

Our study has several limitations. Our study excludes papers not published on ACM and IEEE databases. It does not include research papers about SHD vulnerabilities from other databases that did not appear in the selected databases.

Another limitation is that we have included articles published only in English and not included articles published after March 2019.

Moreover, the literature search keywords were chosen to match the goal of our study as much as possible. However, the search keywords used might have left out articles that included SHV information but did not use the selected keywords.

Finally, many manufacturers and vendors do update their software with patches as soon as a vulnerability is published, and protocols with flaws are updated. Thus, a limitation of our paper is that some vulnerabilities may no longer exist. However, the vulnerabilities information, categorization and taxonomy will serve as a basis for future research in newer types, brands and models of SHDs and their components.

6 Conclusion

We performed a systematic literature review to study 153 SHD vulnerabilities from 98 papers, categorized the vulnerabilities based on their characteristics, and proposed a taxonomy for the attacks. We also discussed solutions to these vulnerabilities from research literature and presented potential opportunities in the area of SHD security and privacy research.

A smart home is a mix of heterogenous devices, controllers, protocols, and software from wide variety of vendors and manufacturers. In addition to the efforts of adding security and privacy tools to SHDs, more research is necessary to design SHDs with security and privacy in mind. A combination of built in security and privacy features with home network protection tools can make mitigation stronger.

Most of the solutions proposed are prototypes, not fully functional tools ready for consumer use. SHD users will benefit from deployable tools. Future work should focus on developing more working solutions that can be made available to the consumers for the protection of the smart home.

Acknowledgment. This research was funded in part by 4-VA, a collaborative partnership for advancing the Commonwealth of Virginia.

References

1. Cost of Data Breach Study (2018). www.ibm.com/security/data-breach
2. Ahmad, W., Sunshine, J., Kaestner, C., Wynne, A.: Enforcing fine-grained security and privacy policies in an ecosystem within an ecosystem. In: Proceedings of the 3rd International Workshop on Mobile Development Lifecycle, MobileDeLi 2015, pp. 28–34. ACM, New York (2015). https://doi.org/10.1145/2846661.2846664
3. Alanwar, A., Balaji, B., Tian, Y., Yang, S., Srivastava, M.: EchoSafe: sonar-based verifiable interaction with intelligent digital agents. In: Proceedings of the 1st ACM Workshop on the Internet of Safe Things, SafeThings 2017, pp. 38–43. ACM, New York (2017). https://doi.org/10.1145/3137003.3137014

4. Alharbi, R., Aspinall, D.: An iot analysis framework: an investigation of IoT smart cameras' vulnerabilities. Living Internet Things: Cybersecur. IoT **2018**, 1–10 (2018)
5. Antonakakis, M., April, T., Bailey, M., Bursztein, E., Cochran, J., Durumeric, Z., Alex Halderman, J., Menscher, D., Seaman, C., Sullivan, N., Thomas, K., Zhou, Y.: Understanding the Mirai botnet. In: Proceedings of the 26th USENIX Security Symposium, Vancouver, BC, Canada, pp. 1093–1110 (2017). https://www.usenix.org/conference/usenixsecurity17/technical-sessions/presentation/antonakakis
6. Anwar, M.N., Nazir, M., Mustafa, K.: Security threats taxonomy: smart-home perspective. In: 2017 3rd International Conference on Advances in Computing, Communication & Automation (ICACCA) (Fall), pp. 1–4 (2017)
7. Aouini, I., Ben Azzouz, L., Jebali, M., Saidane, L.A.: Improvements to the smart energy profile security. In: 2017 13th International Wireless Communications and Mobile Computing Conference (IWCMC), pp. 1356–1361 (June 2017)
8. Apthorpe, N., Reisman, D., Feamster, N.: A smart home is no castle: privacy vulnerabilities of encrypted IoT traffic. In: Data and Algorithmic Transparency Workshop (DAT), New York (2016). http://datworkshop.org/papers/dat16-final37.pdf
9. Apthorpe, N., Reisman, D., Sundaresan, S., Narayanan, A., Feamster, N.: Spying on the smart home: privacy attacks and defenses on encrypted IoT traffic. arXiv Preprint arxiv:1708.05044 (2017)
10. Beyer, S.M., Mullins, B.E., Graham, S.R., Bindewald, J.M.: Pattern-of-life modeling in smart homes. IEEE Internet Things J. **5**(6), 5317–5325 (2018)
11. Braga, M.: People are complaining that Amazon Echo is responding to Ads on TV (2015)
12. Bugeja, J., Jönsson, D., Jacobsson, A.: An investigation of vulnerabilities in smart connected cameras. In: 2018 IEEE International Conference on Pervasive Computing and Communications Workshops (PerCom Workshops), pp. 537–542 (March 2018)
13. Chang, V., Chundury, P., Chetty, M.: "Spiders in the sky": user perceptions of drones, privacy, and security. In: Chi 2017 (2017). https://hci.princeton.edu/wp-content/uploads/sites/459/2017/01/CHI2017_CameraReady.pdf
14. Chatfield, B., Haddad, R.J.: RSSI-based spoofing detection in smart grid IEEE 802.11 home area networks. In: 2017 IEEE Power Energy Society Innovative Smart Grid Technologies Conference (ISGT), pp. 1–5 (April 2017)
15. Chhetri, C.: Towards a smart home usable privacy framework. In: Conference Companion Publication of the 2019 on Computer Supported Cooperative Work and Social Computing, CSCW 2019, pp. 43–46. Association for Computing Machinery, New York (2019). https://doi.org/10.1145/3311957.3361849
16. Chhetri, C., Motti, V.G.: Eliciting privacy concerns for smart home devices from a user centered perspective. In: Taylor, N.G., Christian-Lamb, C., Martin, M.H., Nardi, B. (eds.) Information in Contemporary Society, pp. 91–101. Springer, Cham (2019). https://doi.org/10.1007/978-3-030-15742-5_8
17. Cipriani, J.: What you need to know about encryption on your phone (2016). https://www.cnet.com/news/iphone-android-encryption-fbi/
18. Conti, M., Kaliyar, P., Rabbani, M.M., Ranise, S.: Split: a secure and scalable RPL routing protocol for Internet of Things. In: 2018 14th International Conference on Wireless and Mobile Computing, Networking and Communications (WiMob), pp. 1–8 (2018)
19. Crossman, M.A., Hong, L.: Study of authentication with IoT testbed. In: 2015 IEEE International Symposium on Technologies for Homeland Security (HST), pp. 1–7 (April 2015)

20. Das, A.K., Zeadally, S., Wazid, M.: Lightweight authentication protocols for wearable devices. Comput. Electr. Eng. **63**, 1–13 (2017). http://linkinghub.elsevier.com/retrieve/pii/S0045790617305347
21. Fan, K., Wang, S., Ren, Y., Yang, K., Yan, Z., Li, H., Yang, Y.: Blockchain-based secure time protection scheme in IoT. IEEE Internet Things J. **6**, 4671–4679 (2019)
22. Feng, X., Ye, M., Swaminathan, V., Wei, S.: Towards the security of motion detection-based video surveillance on IoT devices. In: Proceedings of the on Thematic Workshops of ACM Multimedia 2017, Thematic Workshops 2017, pp. 228–235. ACM, New York (2017). https://doi.org/10.1145/3126686.3126713
23. Fernandes, E., Jung, J., Prakash, A.: Security analysis of emerging smart home applications. In: 2016 IEEE Symposium on Security and Privacy (SP), pp. 636–654 (2016)
24. Fouda, M.M., Fadlullah, Z.M., Kato, N.: Assessing attack threat against Zigbee-based home area network for smart grid communications. In: The 2010 International Conference on Computer Engineering Systems, pp. 245–250 (November 2010)
25. Fuller, J.D., Ramsey, B.W.: Rogue z-wave controllers: a persistent attack channel. In: 2015 IEEE 40th Local Computer Networks Conference Workshops (LCN Workshops), pp. 734–741 (October 2015)
26. Gawade, A.U., Shekokar, N.M.: Lightweight secure RPL: a need in IoT. In: 2017 International Conference on Information Technology (ICIT), pp. 214–219 (December 2017)
27. Gong, S., Li, H.: Anybody home? Keeping user presence privacy for advanced metering in future smart grid. In: 2011 IEEE GLOBECOM Workshops (GC Wkshps), pp. 1211–1215 (December 2011)
28. Gyory, N., Chuah, M.: IoTOne: integrated platform for heterogeneous IoT devices. In: 2017 International Conference on Computing, Networking and Communications (ICNC), pp. 783–787 (January 2017)
29. Hill, K.: How a creep hacked a baby monitor to say lewd things to a 2-year-old. Forbes.com (2013)
30. Hoenkamp, R., Huitema, G.B., de Moor-van Vugt, A.J.C.: The neglected consumer: the case of the smart meter rollout in the Netherlands. Renew. Energy Law Policy **4**(2011), 269–282 (2014)
31. Hsieh, W., Leu, J.: A dynamic identity user authentication scheme in wireless sensor networks. In: 2013 9th International Wireless Communications and Mobile Computing Conference (IWCMC), pp. 1132–1137 (July 2013)
32. Hung, M.: Leading the IoT. Gartner Inc., Stamford (2017). https://www.gartner.com/imagesrv/books/iot/iotEbook_digital.pdf
33. De Jesus Martins, R., Schaurich, V.G., Knob, L.A.D., Wickboldt, J.A., Filho, A.S., Granville, L.Z., Pias, M.: Performance analysis of 6LoWPAN and CoAP for secure communications in smart homes. In: 2016 IEEE 30th International Conference on Advanced Information Networking and Applications (AINA), pp. 1027–1034 (March 2016)
34. Jia, X., Li, X., Gao, Y.: A novel semi-automatic vulnerability detection system for smart home. In: Proceedings of the International Conference on Big Data and Internet of Thing, BDIOT2017, pp. 195–199. ACM, New York (2017). https://doi.org/10.1145/3175684.3175718
35. Jia, Y., Xiao, Y., Yu, J., Cheng, X., Liang, Z., Wan, Z.: A novel graph-based mechanism for identifying traffic vulnerabilities in smart home IoT. In: IEEE INFOCOM 2018 - IEEE Conference on Computer Communications, pp. 1493–1501 (April 2018)

36. Johnson, R., Elsabagh, M., Stavrou, A., Offutt, J.: Dazed droids: a longitudinal study of android inter-app vulnerabilities. In: Proceedings of the 2018 on Asia Conference on Computer and Communications Security, ASIACCS 2018, pp. 777–791. ACM, New York (2018). https://doi.org/10.1145/3196494.3196549
37. Kolias, C., Kambourakis, G., Stavrou, A., Voas, J.: DDoS in the IoT: Mirai and other botnets. Computer **50**(7), 80–84 (2017)
38. Lau, J., Zimmerman, B., Schaub, F.: Alexa, are you listening? Privacy perceptions, concerns and privacy-seeking behaviors with smart speakers. Proc. ACM Hum.-Comput. Interact. **2**, 102:1–102:31 (2018)
39. Lee, C., Zappaterra, L., Choi, K., Choi, H.-A.: Securing smart home: technologies, security challenges, and security requirements. In: 2014 IEEE Conference on Communications and Network Security, pp. 67–72 (October 2014)
40. Lei, M., Yang, Y., Ma, N., Sun, H., Zhou, C., Ma, M.: Dynamically enabled defense effectiveness evaluation of a home Internet based on vulnerability analysis and attack layer measurement. Pers. Ubiquit. Comput. **22**(1), 153–162 (2018). https://doi.org/10.1007/s00779-017-1084-3
41. Lei, X., Tu, G., Liu, A.X., Li, C., Xie, T.: The insecurity of home digital voice assistants - vulnerabilities, attacks and countermeasures. In: 2018 IEEE Conference on Communications and Network Security (CNS), pp. 1–9 (May 2018)
42. Li, H., Gong, S., Lai, L., Han, Z., Qiu, R.C., Yang, D.: Efficient and secure wireless communications for advanced metering infrastructure in smart grids. IEEE Trans. Smart Grid **3**(3), 1540–1551 (2012)
43. Ling, Z., Luo, J., Xu, Y., Gao, C., Wu, K., Fu, X.: Security vulnerabilities of Internet of Things: a case study of the smart plug system. IEEE Internet Things J. **4**(6), 1899–1909 (2017)
44. Liu, H., Li, C., Jin, X., Li, J., Zhang, Y., Gu, D.: Smart solution, poor protection: an empirical study of security and privacy issues in developing and deploying smart home devices. In: Proceedings of the 2017 Workshop on Internet of Things Security and Privacy, IoTS&P 2017, pp. 13–18. ACM, New York (2017). https://doi.org/10.1145/3139937.3139948
45. Liu, Y., Hu, S., Ho, T.: Leveraging strategic detection techniques for smart home pricing cyberattacks. IEEE Trans. Dependable Secur. Comput. **13**(2), 220–235 (2016)
46. Liu, Y., Hu, S., Ho, T.Y.: Vulnerability assessment and defense technology for smart home cybersecurity considering pricing cyberattacks. In: Proceedings of the 2014 IEEE/ACM International Conference on Computer-Aided Design, ICCAD 2014, pp. 183–190. IEEE Press, Piscataway (2014). http://dl.acm.org/citation.cfm?id=2691365.2691404
47. Liu, Y., Hu, S., Wu, J., Shi, Y., Jin, Y., Hu, Y., Li, X.: Impact assessment of net metering on smart home cyberattack detection. In: Proceedings of the 52nd Annual Design Automation Conference, DAC 2015, pp. 97:1–97:6. ACM, New York (2015). https://doi.org/10.1145/2744769.2747930
48. Ma, X., Goonawardene, N., Tan, H.P.: Identifying elderly with poor sleep quality using unobtrusive in-home sensors for early intervention. In: Proceedings of the 4th EAI International Conference on Smart Objects and Technologies for Social Good, Goodtechs 2018, pp. 94–99. ACM, New York (2018). https://doi.org/10.1145/3284869.3284894
49. Mahadewa, K.T., Wang, K., Bai, G., Shi, L., Dong, J.S., Liang, Z.: Homescan: scrutinizing implementations of smart home integrations. In: 2018 23rd International Conference on Engineering of Complex Computer Systems (ICECCS), pp. 21–30 (December 2018)

50. Malik, K.M., Malik, H., Baumann, R.: Towards vulnerability analysis of voice-driven interfaces and countermeasures for replay attacks. In: 2019 IEEE Conference on Multimedia Information Processing and Retrieval (MIPR), pp. 523–528 (March 2019)
51. McMahon, E., Patton, M., Samtani, S., Chen, H.: Benchmarking vulnerability assessment tools for enhanced cyber-physical system (CPS) resiliency. In: 2018 IEEE International Conference on Intelligence and Security Informatics (ISI), pp. 100–105 (November 2018)
52. Meng, Y., Wang, Z., Zhang, W., Wu, P., Zhu, H., Liang, X., Liu, Y.: Wivo: enhancing the security of voice control system via wireless signal in IoT environment. In: Proceedings of the Eighteenth ACM International Symposium on Mobile Ad Hoc Networking and Computing, Mobihoc 2018, pp. 81–90. ACM, New York (2018). https://doi.org/10.1145/3209582.3209591
53. Menon, D.M., Radhika, N.: Anomaly detection in smart grid traffic data for home area network. In: 2016 International Conference on Circuit, Power and Computing Technologies (ICCPCT), pp. 1–4 (March 2016)
54. Moody, M., Hunter, A.: Exploiting known vulnerabilities of a smart thermostat. In: 2016 14th Annual Conference on Privacy, Security and Trust (PST), pp. 50–53 (December 2016)
55. Mosenia, A., Jha, N.K.: A comprehensive study of security of Internet-of-Things. IEEE Trans. Emerg. Top. Comput. **5**(4), 586–602 (2016)
56. Namboodiri, V., Aravinthan, V., Mohapatra, S.N., Karimi, B., Jewell, W.: Toward a secure wireless-based home area network for metering in smart grids. IEEE Syst. J. **8**(2), 509–520 (2014)
57. Roux, J., Alata, E., Auriol, G., Kaâniche, M., Nicomette, V., Cayre, R.: RadIoT: radio communications intrusion detection for IoT - a protocol independent approach. In: 2018 IEEE 17th International Symposium on Network Computing and Applications (NCA), pp. 1–8 (November 2018)
58. Roux, J., Alata, E., Auriol, G., Nicomette, V., Kâaniche, M.: Toward an intrusion detection approach for IoT based on radio communications profiling. In: 2017 13th European Dependable Computing Conference (EDCC), pp. 147–150 (September 2017)
59. Salami, S.A., Baek, J., Salah, K., Damiani, E.: Lightweight encryption for smart home. In: 2016 11th International Conference on Availability, Reliability and Security (ARES), pp. 382–388 (August 2016)
60. Saleh, M., Al Barghuthi, N.B., Alawadhi, K., Sallal, F., Ferrah, A.: Streamlining "smart grid end point devices" vulnerability testing using single board computer. In: 2018 Advances in Science and Engineering Technology International Conferences (ASET), pp. 1–6 (February 2018)
61. Sanchez, I., Satta, R., Fovino, I.N., Baldini, G., Steri, G., Shaw, D., Ciardulli, A.: Privacy leakages in smart home wireless technologies. In: 2014 International Carnahan Conference on Security Technology (ICCST), pp. 1–6 (October 2014)
62. Shen, T., Ma, M.: Security enhancements on home area networks in smart grids. In: 2016 IEEE Region 10 Conference (TENCON), pp. 2444–2447 (November 2016)
63. Sivanathan, A., Loi, F., Gharakheili, H.H., Sivaraman, V.: Experimental evaluation of cybersecurity threats to the smart-home. In: 2017 IEEE International Conference on Advanced Networks and Telecommunications Systems (ANTS), pp. 1–6 (December 2017)

64. Sivaraman, V., Chan, D., Earl, D., Boreli, R.: Smart-phones attacking smart-homes. In: Proceedings of the 9th ACM Conference on Security & Privacy in Wireless and Mobile Networks, WiSec 2016, pp. 195–200. ACM, New York (2016). https://doi.org/10.1145/2939918.2939925
65. Sun, A., Gong, W., Shea, R., Liu, J.: A castle of glass: leaky IoT appliances in modern smart homes. IEEE Wirel. Commun. **25**(6), 32–37 (2018)
66. Tabrizi, F.M., Pattabiraman, K.: Intrusion detection system for embedded systems. In: Proceedings of the Doctoral Symposium of the 16th International Middleware Conference, Middleware Doct Symposium 2015, pp. 9:1–9:4. ACM, New York (2015). https://doi.org/10.1145/2843966.2843975
67. Tekeoglu, A., Tosun, A.: Blackbox security evaluation of Chromecast network communications. In: 2014 IEEE 33rd International Performance Computing and Communications Conference (IPCCC), pp. 1–2 (December 2014)
68. Tekeoglu, A., Tosun, A.: A closer look into privacy and security of Chromecast multimedia cloud communications. In: 2015 IEEE Conference on Computer Communications Workshops (INFOCOM WKSHPS), pp. 121–126 (April 2015)
69. Trimananda, R., Younis, A., Wang, B., Xu, B., Demsky, B., Xu, G.: Vigilia: securing smart home edge computing. In: 2018 IEEE/ACM Symposium on Edge Computing (SEC), pp. 74–89 (October 2018)
70. Whitehurst, L.N., Andel, T.R., McDonald, J.T.: Exploring security in ZigBee networks. In: Proceedings of the 9th Annual Cyber and Information Security Research Conference, CISR 2014, pp. 25–28. ACM, New York (2014). https://doi.org/10.1145/2602087.2602090
71. Wurm, J., Hoang, K., Arias, O., Sadeghi, A., Jin, Y.: Security analysis on consumer and industrial IoT devices. In: 2016 21st Asia and South Pacific Design Automation Conference (ASP-DAC), pp. 519–524 (January 2016)
72. Zeng, E., Mare, S., Roesner, F.: End user security & privacy concerns with smart homes. In: Symposium on Usable Privacy and Security (SOUPS) (2017)
73. Zhang, M., Liu, Y., Wang, J., Hu, Y.: A new approach to security analysis of wireless sensor networks for smart home systems. In: 2016 International Conference on Intelligent Networking and Collaborative Systems (INCoS), pp. 318–323 (September 2016)

Short Papers

Information Warfare and Cyber Education: Issues and Recommendations

Joshua A. Sipper(✉)

Air University, Air Force Cyber College, Montgomery, AL 36112, USA
jasipper@gmail.com

Abstract. With the introduction of the Information Warfare (IW) paradigm across military service components, the institution of a new education focus is apropos. The purpose of this paper is to discuss the shifts in thought regarding IW and how the introduction of the fields of intelligence, surveillance, and reconnaissance (ISR), electromagnetic warfare (EW), and information operations (IO) is changing how we educate our soldiers, sailors, airmen, marines, and government civilians. The multidisciplinary ramifications will be discussed separately in order to better define each discipline's requirements and individual foci as they apply to cyber operations (CO), ISR, EW, and IO.

Keywords: Cyber · Education · Information · Warfare · Intelligence

1 Introduction

The word education carries with it several interesting connotations: professional, strategic, non-technical, leadership, and managerial, to name a few. "The [Air Force Association CyberPatriot] program was well received by industry professionals and is now sponsored by multiple corporations including Northrop Grumman Foundation, Cisco, Symantec, and the University of Maryland University College" (Dill 2018). Other organizations are taking technical understanding to a more fundamental level. The United States Military Academy (USMA) Mathematical Sciences Department is using mathematics education to "help prepare future military officers for leadership roles in the cyber-affected world in three tiers: (1) what all officers should know, (2) what highly technical officers should know, and (3) what cyber leaders should know" (Arney et al. 2016). This highly technical focus belies the fact that cyber is a career field that not only crosses boundaries, but one filled with progressive challenges. With these challenges comes the need to educate personnel to a standard that includes a prismatic display of knowledge, skills, and abilities (KSAs). Education is certainly a key to staffing and operating within cyberspace. Incorporating technology, advanced educational strategies, and technically focused education all play a role in assembling the Joint All Domain Operations (JADO) force necessary to ensure the nation's security and dominance across the global cyber cosmos.

K.-K. R. Choo et al. (Eds.): NCS 2020, AISC 1271, pp. 235–237, 2021.
https://doi.org/10.1007/978-3-030-58703-1_14

2 Cyber and IW

CO and the education undergirding this capability has been given a new mandate: interoperability with the ISR, EW, and IO disciplines. It is with this new perspective in mind that educators must proceed, bringing with them the responsibility of interleaving this panoply of IW capabilities and disciplines. "The use of cyber assets has been a form of force projection that helps initiate crises far ahead of and beyond the frontlines… new technologies have been able to provide both the means and vulnerabilities to allow such operations at a scale not often witnessed before, and with a smaller investment in resources on the part of the aggressor" (Danyk et al. 2017). It is clear that cyber has wormed its way into basically every area of life and shows no sign of stopping and has also been established as a domain, specific to its own capabilities and effects. ISR as a capability is also maturing across the globe. "Foreign intelligence services use cyber tools in information-gathering and espionage. Several nations are aggressively working to develop information warfare doctrine, programs, and capabilities to enable a single entity to have a significant and serious impact by disrupting the supply, communications, and economic infrastructures that support military power" (Jabbour and Devendorf 2017). With this in mind, it is important to see the advantages of such constructs and how to meet the challenges of other nation states and the capabilities they continue to develop. "[EW] can … be carried out by controlling devices that emit radio-frequency (RF) energy. New forms of RF signals pervade homes and cities: Bluetooth, Wi-Fi, 5G, keyless entry systems, and Global Positioning System (GPS), to name a few. The coming Internet of Things (IoT) is essentially an Internet of RF-connected items." (Libicki 2017). With this powerful reach into the RF spectrum, EW stands as an excellent, cyber enabled resource, capable of combining with other IRCs in many, powerful ways. "[I]nformation operations offers an opportunity to achieve a level of dominance… it provides a significantly less costly method of conducting operations since it replaces the need for conventional military forces" (Ajir and Vailliant 2018). It is difficult not to see how powerful IO is in regards to influence and dominance since information has become and remains a key to everything from business to commerce to military operations, especially as it relates to social media.

3 Recommendations

While some institutions have already begun to delve into interdisciplinary education regarding the IW IRCs, the integration of education regarding these capabilities and their interoperability must be further explored. This can be done through the introduction of curricula in a cross-disciplinary fashion to familiarize students with each capability while keeping their own discipline at the forefront allowing them the focus they need and introducing them to how cyber and the other IRCs operate within the larger IW construct. Also, early exposure to the actual operational IW environment could be of special significance to students as they can get a first-hand look at how these IRCs interleave and fuse together into a holistic product.

Disclaimers. Opinions, conclusions, and recommendations expressed or implied within are solely those of the author and do not necessarily represent the views of the Air University, the United States Air Force, the Department of Defense, or any other US government agency.

DoD School Policy. DoD gives its personnel in its school environments the widest latitude to express their views. To ensure a climate of academic freedom and to encourage intellectual expression, students and faculty members of an academy, college, university, or DoD school are not required to submit papers or material that are prepared in response to academic requirements and not intended for release outside the academic institution. Information proposed for public release or made available in libraries or databases or on web sites to which the public has access shall be submitted for review.

This is a shortened version of the paper titled "It's Not Just About Cyber Anymore: Multidisciplinary Cyber Education and Training under the New Information Warfare Paradigm" originally published in Joint Forces Quarterly. Express permission for this condensed version granted by JFQ.

References

Ajir, M., Vailliant, B.: Russian information warfare: implications for deterrence theory. Strat. Stud. Q. **12**(3), 70–89 (2018)

Arney, C., Vanatta, N., Nelson, T.: Cyber education via mathematical education. Cyber Def. Rev. **1**(2), 49–60 (2016)

Danyk, Y., Maliarchuk, T., Briggs, C.: Hybrid WAR: high-tech, information and cyber conflicts. Connections **16**(2), 5–24 (2017)

Dill, K.: Cybersecurity for the nation: workforce development. Cyber Def. Rev. **3**(2), 55–64 (2018)

Jabbour, K.T., Devendorf, E.: Cyber threat characterization. Cyber Def. Rev. **2**(3), 79–94 (2017)

Libicki, M.C.: The convergence of information warfare. Strat. Stud. Q. **11**(1), 49–65 (2017)

Designing an Internet-of-Things Laboratory to Improve Student Understanding of Secure Embedded Systems

A. R. Rao(✉), Kavita Mishra, and Nagasravani Recharla

Fairleigh Dickinson University, Teaneck, NJ 07666, USA
raviraodr@gmail.com, {kavita02,nagasravani}@student.fdu.edu

Abstract. In the U.S., the Department of Defense and the National Security Agency are taking steps to build cybersecurity capacity through workforce training and education. We describe the development of cybersecurity education courseware for internet-of-things (IoT) applications.

We selected the domain of healthcare and featured different IoT sensors that are seeing increased usage. These include barcode scanners, cameras, fingerprint sensors, and pulse sensors. These devices cover important functions such as patient identification, monitoring, and creation of electronic health records. We used a password protected MySQL database as a model for electronic health records. We demonstrated potential vulnerabilities of these databases to SQL injection attacks.

We administered these labs during a Fall 2019 course where students reported a significant increase in their understanding of cybersecurity issues. The instructional lab material will be uploaded to https://clark.center.

Keywords: Cybersecurity · Curriculum development · Internet of Things

1 Introduction

In an earlier paper at the 2019 National Cyber Summit [1], we created a framework to integrate cybersecurity and its associated learning modules into engineering programs especially in the area of embedded system design. We used hands-on lab exercises where students learnt configure multiple IoT devices while maintaining a "security mindset." We emphasized the development of hands-on laboratory exercises as our previous research demonstrated their ability to excite and engage students [2–5].

2 Methods and Results

We selected the Raspberry-Pi platform, with the addition of several inexpensive sensors that can model the creation and use of healthcare data. We implemented and tested several lab exercises to teach students about secure embedded systems in healthcare and retail applications. In order to give the students a real-world IoT experience, we used the following equipment: a WoneNice USB Laser Barcode Scanner (wired) [6], a fingerprint

K.-K. R. Choo et al. (Eds.): NCS 2020, AISC 1271, pp. 238–239, 2021.
https://doi.org/10.1007/978-3-030-58703-1_15

reader manufactured by Zhian Tec, and a pulse sensor made by pulsesensor.com. Students understood concepts related to data acquisition, storage through MySQL databases, and protection. They studied vulnerabilities in database systems, which could be subject to SQL injection attacks.

3 Discussion and Conclusion

A well-designed embedded systems course can contain several such labs to stimulate students and improve their knowledge of important hardware and software components in the ever-expanding technology ecosystem. A well-trained embedded systems engineer will be conversant in both hardware and software technologies, and this combination of skills will be important in securing cyber-physical systems of the future.

Students reported that their knowledge of cybersecurity issues improved significantly after completing the lab exercises related to secure embedded systems. The mean confidence level in cybersecurity issues increased from 2.5 to 4.1 on a 5-point scale after taking this course. Students were definitely eager and excited to learn a new perspective of cybersecurity that they did not obtain in other courses.

Acknowledgment. We are grateful to the students in the Fall 2019 Embedded Systems course at Fairleigh Dickinson University. The project was sponsored by the National Security Agency under Grant/Cooperative Agreement Fairleigh Dickinson University CySP Grant Number H98230-19-1-0272. The United States Government is authorized to reproduce and distribute reprints notwithstanding any copyright notation herein.

References

1. Rao, A.R., Clarke, D.: Capacity building for a cybersecurity workforce through hands-on labs for Internet-of-Things security. In: National Cyber Summit, pp. 14–29 (2019)
2. Rao, A.R.: Interventions for promoting student engagement and predicting performance in an introductory engineering class. In: Advances in Engineering Education (2020)
3. Rao, A.R., Dave, R.: Developing hands-on laboratory exercises for teaching STEM students the Internet-of-Things, cloud computing and blockchain applications. In: IEEE Integrated STEM Education Conference, Princeton, NJ (2019)
4. Rao, A.R., Clarke, D., Bhadiyadra, M., Phadke, S.: Development of an embedded system course to teach the Internet-of-Things. In: IEEE STEM Education Conference, ISEC, Princeton, pp. 154–160 (2018)
5. Rao, A.R.: A three-year retrospective on offering an embedded systems course with a focus on cybersecurity. In: IEEE STEM Education Conference, ISEC-2020, Princeton, NJ (2020)
6. https://www.amazon.com/gp/product/B00LE5VV1C/ref=as_li_ss_tl

Distributed Denial of Service Attack Detection

Travis Blue and Hossain Shahriar(✉)

Department of Information Technology, Kennesaw State University, Kennesaw, USA
tblue5@students.kennesaw.edu, hshahria@kennesaw.edu

Abstract. Distributed Denial of Service (DDoS) attacks has been a persistent threat for network and applications. Successful attacks can lead to inaccessible service to legitimate users in time and loss of business reputation. In this paper, we explore DDoS attack detection using Term Frequency (TF)-Inverse Document Frequency (IDF) and Latent Semantic Indexing (LSI). We analyzed web server log data generated in a distributed environment.

Keywords: Denial of Service · Latent Semantic Index · Term-Frequency · Inverse Document Frequency

1 Introduction

Distributed Denial of Service (DDoS) attacks occur by issuing a large number of requests to a target web server. Existing network layer approaches are not applicable for detecting App-DDoS attacks. A number of bots are available in the market that can automate application layer DDoS attacks such as Dirtjumper. DDoS as a service is now currently available to mount attacks on legitimate entities [1, 2]. DDoS attacks have been mounted against various websites such as game (Sony Play Station) [3] and bitcoin [4]. App-DDoS attacks can lead to loss in revenue ($5600 per minute), productivity, and reputation [5]. In this paper, we develop a mitigation approach for DDoS.

We first identify page ranking using a popular Term Frequency (TF)-Inverse Document Frequency (IDF) from web server logs. Then, we identify ranking of resources that are accessed most to build normal profile. For a given web session, we form a query of accessed resources and find how close it is with respect to the normal profile using Latent Semantic Indexing (LSI). If a large deviation is identified, the session is identified as part of DDoS attack.

2 TF-IDF and LSI Approach

TF-IDF is a computation approach to find the importance of word in a set of documents. It is composed by computing two terms: TF and IDF as discussed below. Term Frequency (TF) measures how frequently a term occurs in a document.

$$TF(t) = (\#\ of\ \ times\ t\ appears\ in\ a\ document)/(Total\ \#\ of\ \ terms\ in\ the\ document) \ldots\ldots \quad \text{(i)}$$

K.-K. R. Choo et al. (Eds.): NCS 2020, AISC 1271, pp. 240–241, 2021.
https://doi.org/10.1007/978-3-030-58703-1_16

Inverse Document Frequency (IDF) measures how important a term is in all document. TF assumes that each is equally important. However, certain term may be common (e.g., article in sentence). Thus, IDF consider the frequent term as rare and less occurring term as important.

$$IDF(t) = log_e(Total\ \#\ of\ documents / \#\ of\ documents\ having\ t) \ldots\ldots \quad \text{(ii)}$$

We apply Latent Semantic Indexing (LSI) technique to perform query and obtain the close similarity of documents for decision making. We represent access pattern from sample data in a matrix form to apply LSI, where each row represents specific resource access for a certain day, and column represents specific words found in log line with their TF-IDF values obtained from previous step. The columns would not only contain web page name, but also specific resources such as images, amount of bytes, status code, browser name.

For DDoS attack detection, we consider building a query obtained from an ongoing session data. We extract the resources (term) of interests from the log. Log files (documents) obtained for legitimate traffic for various days are used form Term-Document matrix. Terms are defined based on words representing resources (php files, image files) having non-zero TF-IDF value (discussed in previous section). Queries obtained for given sessions are used to find how close a given day's log represent for an ongoing session. Similarity measures are based on cosine metrics. The closer a query vector and a document is, the higher the distance. Hence, if distance is above certain threshold (d) level, it is not considered an attack. If the similarity is less than a threshold value, then the ongoing session is considered as an attack.

References

1. Cheng, J., Yin, J., Liu, Y., Cai, Z., Wu, C.: DDoS attack detection using IP address feature interaction. In: Proceedings of 2009 International Conference on Intelligent Networking and Collaborative Systems, pp. 113–118 (2009)
2. Wueest, C.: The continued rise of DDoS attacks. Symantec Technical Report, October 2014. https://www.symantec.com/content/en/us/enterprise/media/security_response/whitepapers/the-continued-rise-of-ddos-attacks.pdf
3. Sony Playstation Hack (2016). https://www.scmagazine.com/sony-psn-downed-hacking-group-claims-ddos-attack/article/463065/
4. Bienkowski, T.: Your Network or Your Bitcoins: Three Rules for Dealing with DDoS Extortion Threats, February 2016. https://www.arbornetworks.com/blog/insight/your-network-or-your-bitcoins-three-rules-for-dealing-with-ddos-extortion-threats/
5. What is a DDoS Attack? (2020). https://www.verisign.com/en_US/security-services/ddos-protection/what-is-a-ddos-attack/index.xhtml

Is There a Prophet Who Can Predict Software Vulnerabilities?

Michael T. Shrove[1(✉)] and Emil Jovanov[2]

[1] Millennium Corporation, 340 The Bridge Street Suite 212, Huntsville, AL 35806, USA
tshrove@gmail.com
[2] University of Alabama Huntsville, 301 Sparkman Drive, Huntsville, AL 35899, USA
emil.jovanov@uah.edu

Abstract. Shortcuts in software development generate technical debt and software vulnerabilities. We propose a framework that will allow stakeholders an effective way to forecast the trend in software vulnerabilities and allow stakeholders to provide the necessary resources to reduce the attack surface and the probability of software failure. We demonstrated that our method can forecast vulnerabilities in several open-source projects, and seasonality in daily, monthly, and yearly total vulnerabilities. Our preliminary results indicate that we can use forecasting methods up to 90 days out with accuracy. In this paper, we present our technique, methodology of preparation of inputs for the proposed artificial intelligence model, and the results of analysis of three open source projects.

Keywords: Vulnerabilities · Security debt · Machine learning · FBProphet

1 Introduction

In today's world, software has almost become a part of every job function in the market. With the increase in demand for software, software developers are pressured to produce software quickly. Therefore, software developers are often taking shortcuts in code development in order to meet their goals and milestones, which produces technical debt (TD) [1]. In this research, we are interested in the TD costs associated with shortcuts in the security aspects of the software which increases the security debt, also known as attack surface [2].

For this research, our goal was to develop a method that can be used by software teams to monitor their security debt, forecast seasonal changes, and predict when to expect additional software vulnerabilities. We used Facebook's Prophet framework (FBProphet) to forecast trends in security debt, as described in Sect. 2. The proposed method would allow stakeholders to reallocate their resources more effectively and anticipate future releases of security patches.

2 Overall Approach and Results

We used data from three open source projects (Kubernetes, Brave Browser, and Electron) to demonstrate the effectiveness of the proposed approach. GitHub API is used to extract

K.-K. R. Choo et al. (Eds.): NCS 2020, AISC 1271, pp. 242–243, 2021.
https://doi.org/10.1007/978-3-030-58703-1_17

security issues for each project. We decided to calculate the security debt cost as + 1 (production of vulnerability) or −1 (resolving the vulnerability). We arranged the security debt items by date from the earliest dated debt first, to the most recent dated debt last. We then created a cumulative sum of the security debt by the project.

Next, we tested for the stationarity of each project's data using the Dickey-Fuller (DF) test. The results show that all three data sets are nonstationary. Our research showed that the Difference Natural Log (Diff Log) was the best technique to make our data stationary. Once we transformed all projects' data into stationary data, we used the FBProphet model to forecast the data.

Prophet is an open-source forecasting tool developed by Facebook. It is used for forecasting time series data based on an additive model where non-linear trends are fit with yearly, weekly, and daily seasonality. Lastly, we apply the FBProphet model to each dataset after the Diff Log transformation method has been applied. The results showed that using our technique, we resulted in an average mean absolute error (MAE) of 9.53 vulnerabilities or 17.4% for all three projects forecasting out 90 days in the future.

3 Discussion and Conclusion

We demonstrated that the proposed method was effective in prediction of software vulnerabilities. In addition, the method reveals daily, weekly, and yearly seasonal components of vulnerabilities. As an example, the Brave project seems to increasingly produce vulnerabilities starting from the middle of the year until the end of the year, whereas the Electron project had a pattern of producing vulnerabilities from November to December timeframe. With this insight the stakeholders could adjust resources and add additional personnel during critical periods to correct those vulnerabilities. Consequently, they can lower the attack surface in real-time as the security debt arises. Large software organizations could use this research to forecast seasonal trends across multiple projects and schedule one security team to rotate through projects to lower the attack surface based on annual trends.

By providing this mechanism, stakeholders can effectively schedule resources. They could also promote campaigns within the open-source community to solicit additional help. In this research, we have shown that using FBProphet, one can forecast security vulnerabilities with confidence. In the future, we would like to use traditional technical debt along with security debt to produce a three-dimensional model and show how complexity in software can help forecast security debt in a software project.

References

1. Cunningham, W.: The WyCash portfolio management system. ACM SIGPLAN OOPS Messenger **4**(2), 29–30 (1993). https://doi.org/10.1145/157710.157715
2. Campos, M., Silva, O., Valente, M.T., Terra, R.: Does technical debt lead to the rejection of pull requests? In: Brazilian Symposium on Information Systems (SBSI), pp. 1–7 (2016). https://arxiv.org/pdf/1604.01450.pdf

Author Index

K. R. Choo et al. (Eds.): NCS 2020, AISC 1271, pp. 245–246, 2021.
https://doi.org/10.1007/978-3-030-58703-1